S364 EVOLUTION

Unit 12 Species and Speciation
Unit 13 Macroevolution and Development

Prepared by an Open University Course Team

The Open University Press

S364 COURSE TEAM

Chairmen Dennis Jackson (till October 1980) and Irene Ridge

General Editor Irene Ridge

Course Support Alastair Ewing
Mavis Fewtrell

Authors Bernard Campbell (External Consultant)
Simon Conway Morris
Tim Halliday
Mae-Wan Ho
Dennis Jackson
D. Graham Jenkins
David Norman (Consultant), Queen Mary College, London
Caroline Pond
Irene Ridge
Bob Savage (Consultant), University of Bristol
Jonathan Silvertown
Peter Skelton
Charles Turner

Editors Janet Evans
Sue Walker

Other Members Mary Bell (Staff Tutor)
Eric Bowers (Staff Tutor)
Peter Firstbrook (BBC)
Jenny Hill (Cartographer)
Barbara Hodgson (IET)
Stephen Hurry
Mike Johnstone (BBC)
Roger R. Jones (BBC)
Patrick Murphy
John A. Taylor (Illustrator)
Tag Taylor (Designer)
Peggy Varley

UNIT 12   SPECIES AND SPECIATION

Contents

TABLE A1 List of terms and concepts assumed from the Science Foundation Courses*

Term	S100 Unit No.	S101 Unit No.	Term	S100 Unit No.	S101 Unit No.
anaphase	—	25, AV	homozygous	19	19
assortment of chromosomes	19	19	linkage	17, 19	19
centromere	—	25	meiosis	17	25, AV
chromosome	17	19, 25	metaphase	—	25, AV
DNA	13, 17	25	mitosis	17, 19	25, AV
F_1 and F_2 hybrids	—	19	protein	13	16/17, 23
gene	17	19	recombination	19	19
heterozygous	19	19	segregation	—	19
homologous chromosome	—	19, 25	trophic level	20	21

* The Open University (1971) S100 *Science: A Foundation Course*, The Open University Press; or The Open University (1979) S101 *Science: A Foundation Course*, The Open University Press.

TABLE A2 List of terms and concepts from previous Units† and/or developed in this Unit or in the Glossary‡

Term	Section No. in this Unit	Term	Section No. in this Unit	Term	Section No. in this Unit
acrocentric chromosomes‡	4.1	*genetic distance, D‡ (8)	2.1.2, 3.2.2, 4.2.1	niche‡ (1, 11)	4.4
adaptive trend (8)†	5.2.1			non-complementary	
allele‡ (10)	2.1.2, 3.2.2	genetic divergence	2.2, 2.3, 5.1	nucleotides	4.2.2
allopatric speciation‡ (11)	5.1, 6	genetic drift‡ (10)	3.1.3, 5.1.2	organismal differences	4.3
*allopolyploid	4.1	*genetic identity, I	2.1.2, 3.2.2	parthenogenesis (11)	6
aneuploidy‡	4.1	*genetic revolution	5.1.2	*peripheral isolates	3.1.3, 5.1
anther (7)	4.1.2	(reorganization)		*polygenes (10, 11)	3.2.1
antibody	4.2.3	genome‡ (10)	4	polymorphism (1, 10)	4.1
antigen	4.2.3	*geographic isolate	3.1.3	*polyploidy‡	4.1
*autopolyploid	4.1	geographic race‡	3.1	*polytene	
		*geographical isolation	5, 5.1.1	chromosomes‡	4.1.1, AV 8
Bergmann's rule	3.1.2	glacial refuge	3.1.4	*polytypic species‡	3.1.1, 3.2, 4.1.2
centromere‡ (2)	4.1	gradual genetic change	3.2.1		
character		gradualist evolution (8)	6.2	*post-mating (post-zygotic)	
displacement‡ (11)	5.1.3	*Haplochromis*	5.2.1	isolating mechanisms	2.2, 2.2.2, 4, 5.1.3
*chromosomal inversion	4.1, 4.1.1	*Heliconius*, mimicry in	3.2.1		
*chromosomal morphology	4.1	homology‡ (1)	4.2.2	*pre-mating (pre-zygotic)	
*chromosomal mutations‡	4.1	'hopeful monster'	4	isolating mechanisms	2.2, 2.2.1, 5.1.3
*chromosomal		*hybrid breakdown	2.2, 2.2.2	punctuated	
translocation	4.1	*hybrid inviability	2.2, 2.2.2	equilibrium‡ (8)	6.2
cichlid fishes in Lake		*hybrid sterility	2.2, 2.2.2	*regulatory genes‡	4.3
Victoria	5.2.1	*hybrid zones	3.1.4	reproductive gap	4
Clarkia	4.1.2			*reproductive isolation	2.1.3, 5.1, 5.1.3
*clinal variation	3.1.2	immunological			
cline‡	3.1.2	distance‡	4.2.3	*saltational speciation	3.2.1, 4.1.2
cnidarians‡ (3)	2.1.1	*interspecific variation	4	saltatory genetic change	3.2.1
co-adaptation of genes	4	*intraspecific variation	3.1	semispecies	3.1.1, 3.2.2
denaturation of DNA	4.2.2	*intrinsic factors of		sex chromosomes	4.1
*dendrograms‡	3.2.2	geographic isolation	5.1.1	*sibling species‡ (1, 11)	2.2.1, 3.2.2
diapause	6.1	*isolating mechanisms‡	2.2	somatic cells	4.1
disruptive selection (10)	6.1			*species definition (1)	2
DNA hybridization‡	4.2.2	*karyotype‡	4.1	*species flocks	5.2
*ecological shift (11)	5.1.2	*karyotype evolution	4.1.2, 4.1.3	*species swarms	5.2
electrophoresis (10)	2.1.2, 3.2.2, 4.2.1	locus (10)	2.1.2	stabilizing selection‡ (10)	3.2.1
				stigma (7)	4.1.2
*extrinsic factors of		*macrogeographic race	3.1.1	structural genes‡	4.3
geographic isolation	5.1.1	*Mayr's model for		*subspecies	3.1.1, 3.2.2
*fission of chromosomes	4.1	allopatric speciation	5.1	supergene (10)	4.1
fixation of genes/alleles (10)	4.1.3, 5.1.2	metacentric		*sympatric speciation‡ (11)	5.1.1, 6
founder effect (10)	3.1.3, 4.1.3, 5.1.2	chromosome‡	4.1	*systemic mutations	4
		*microgeographic race	3.1.1	teleocentric chromosomes‡	4.1
*fusion of chromosomes	4.1	*monotypic species	3.1.1	trends (5, 6, 8)	5.2.1
gene exchange	3.1.3	morphological		trophic specialization	4.4, 5.2.1
gene flow‡ (10)	3.1.1–3.1.3	discontinuities	4	*vagility	4.1.3, 6
*genetic differences between		mosaic evolution‡ (8)	4.2.3		
species	4.2, 4.3	*Mullerian mimicry	3.2.1		

* The most important terms are indicated by an asterisk.
† The number of the previous Unit is given in parentheses, after the term.
‡ Items followed by a double dagger are to be found in the Glossary (Section 4 in the Handbook).

Objectives

When you have completed this Unit you should be able to:

1 Define and recognize the best definitions of the terms, concepts and principles marked with an asterisk in Table A2 and distinguish between true and false statements concerning these terms. (SAQ 1)

2 Evaluate the different criteria for recognizing species. (SAQ 1)

3 Describe and recognize species isolating mechanisms. (SAQ 2)

4 Recognize difficulties in applying the species concept. (SAQ 1)

5 Describe the typical pattern of variation in a species. (SAQ 3)

6 Evaluate the significance of physical and biotic factors in producing evolutionary change. (SAQ 3)

7 Construct and interpret dendrograms from data on genetic distance. (SAQ 4)

8 Reconstruct phylogenies from data about chromosomal mutations. (SAQ 5)

9 Describe the relationship between the rate of chromosomal evolution and the rate of speciation in mammals and other vertebrates. (SAQ 6)

10 State the relationship between genetic change and organismal evolution. (SAQ 6)

11 State the elements of the model of allopatric speciation and contrast the views of Mayr and Dobzhansky about the origin of reproductive isolation. (SAQ 7)

12 State the 'intrinsic' factors influencing the effectiveness of geographic isolation. (SAQ 7)

13 Contrast the allopatric and sympatric theories on the origin of reproductive isolation mechanisms. (SAQs 7 and 8)

14 Evaluate the significance of the different factors involved in speciation in particular cases. (SAQ 7)

15 Distinguish between allopatric and sympatric speciation. (SAQs 7 and 8)

Study Guide for Unit 12

This Unit examines the species concept (Sections 1 and 2), the nature of intraspecific and interspecific variation (Sections 3 and 4) and the mechanisms of speciation—how new species may arise (Sections 5 and 6). Section 4 is the longest one and you should allow more study time for it than for any other Section. Sections 5 and 6 should be studied with particular care, although they are not very long. If very short of time you could omit the mathematical derivation of genetic distance in Section 2.1.2.

The television programme, 'The picture wings of Hawaii', and AV 8, 'Reconstructing picture wing evolution', are linked with this Unit and are especially relevant to Sections 3 and 4. Both are about evolution and speciation of *Drosophila* in Hawaii. An earlier television programme, 'The adaptive radiation of the silversword alliance', is also relevant to this Unit (mainly to Section 3); it contains material about the role of hybridization in the evolution of Hawaiian plants.

Finally, this Unit is particularly relevant to the Projects about natural selection in plant populations (computer simulation) and about hybridization in plants. Work on these Projects should add to your understanding of the Unit, but it must be emphasized that *you should not write either the reports or the Project essay (if you choose to do this) before reading Unit 12.*

1 Introduction

Ernst Mayr, in 1957, stated:

> Species are important because they represent an important level of integration in living nature. This recognition is fundamental to pure biology, no less than to all subdivisions of applied biology. An inventory of the species of animals and plants of the world is the baseline of further research in biology.

Unfortunately, however, there are both philosophical and practical difficulties associated with species recognition.

When Linnaeus published his monumental work the *Systema Naturae* in 1735, the species was, in his mind, the unit of direct creation—a fixed and objective entity. In opposition to this view, Darwin believed that the species was an arbitrary and subjective entity invented for the purpose of classification by the naturalist. The years following the publication of the *Origin of Species* in 1859 saw biologists clearly divided into two camps: the followers of Linnaeus, and the followers of Darwin. To some extent this division continues to the present day.

If Darwin's intention in writing his book was to explain the origin of species, it may be said that he neatly side-stepped the issue by denying that species existed as separate entities. He believed that there was a continuous gradation of form in nature, expressed in the famous aphorism *natura non facit saltum* (nature does not make a jump).

The study of natural populations themselves, however, suggests that the fauna of a region is composed of units which are often morphologically distinct. Most biologists today implicitly or explicitly accept the existence of species in one form or another. As we shall see, many species do appear to be quite distinct on a number of criteria.

2 The biological concept of species

Study comment This Section examines the species concept, the criteria for distinguishing between species and the isolating mechanisms that keep species apart. The mathematical derivation of genetic distance (Section 2.1.2) need not be memorized, but you should read through it so that you know what genetic distance is.

J. B. Lamarck (1809) produced one of the best early definitions of a species, embodying both the distinctiveness and the mutability required by evolution:

> A species is a collection of similar individuals which are perpetuated by generation in the same condition as long as their environment does not change sufficiently to bring about variation in their habits, their character, and their form.

This definition was the beginning of the modern biological concept of species, for it emphasizes the existence of a biological bond between individuals who are similar in habit, character and form. Plate (1914) was apparently the first to state explicitly the nature of the biological bond:

> The members of a species are tied together by the fact that *they recognise each other as belonging together, and reproduce only with each other*. The systematic category of species is therefore entirely independent of the existence of Man.

The work of field naturalists soon revealed that species typically extend over space and time and, more often than not, possess complex population structures. With the rise of population genetics the biological concept of species finally emerged, which stresses the fact that species are populations of individuals constituting a *reproductive* community that is isolated from other such communities. Accordingly, a species may be aptly defined as *a group of actually or potentially interbreeding natural populations that are reproductively isolated from other such groups* (Unit 1).

definition of species

ITQ 1 On the above definition of species, are the different breeds of dog all members of a single species?

2.1 The criteria for recognition of species

Broadly speaking, three kinds of criteria are used in recognizing species; we shall examine each of them in turn.

2.1.1 Phenotypic criteria

Morphological similarities between individuals of the same species, and their differences from individuals of other species, are the features most often used in classification. Biochemical, physiological and ecological features, when known, are also frequently very valuable in distinguishing one species from another.

☐ What pitfalls are involved when morphological features alone are used in classification?

■ Male and female individuals of sexually dimorphic species (including fossil ones such as the ammonites in Figure 7 of Unit 1 and in the Home Experiment) would be classified in different species, and the confusion becomes even greater in populations exhibiting polymorphism (Unit 10). Yet another pitfall is in organisms with complex life histories (such as cnidarians and many insects), where different developmental stages may be classified as different species.

So, morphology by itself can be quite misleading. There are situations, however, where only morphological data are available, e.g. in the classification of fossils (see Unit 1). The criterion of reproductive isolation obviously cannot be applied to asexual organisms. Such species are exceptions and must be classified using morphological as well as physiological and ecological data.

2.1.2 Genetic criteria

If species are groups of interbreeding (or potentially interbreeding) populations, individuals of the same species should be genetically more similar to one another than to individuals of other species. However, we have seen in Unit 10 that even individuals within a single population can be genetically quite diverse, so we need a measure of genetic similarity that takes this into account.

Measures of genetic distance

There are a number of methods for estimating genetic dissimilarity between populations. The one most often followed, and followed here, is that of M. Nei:

Consider two random-mating diploid populations X and Y in which multiple alleles segregate at a locus. Let the alleles be $A_1, A_2, \ldots A_m$, and let the frequencies of these alleles in populations X and Y be $x_1, x_2, \ldots x_m$, and $y_1, y_2, \ldots y_m$, respectively. The probability (j_x) that two randomly chosen alleles are identical in population X is given by

$$j_x = x_1^2 + x_2^2 + \ldots + x_m^2$$

or

$$j_x = \sum_{i=1}^{m} x_i^2$$

where $x_i = x_1, x_2, \ldots x_m$ and $\sum_{i=1}^{m}$ is a summation sign indicating the summing together of all the individual x_i^2 values from $x_i = x_1$ to $x_i = x_m$. Similarly, in population Y, the probability (j_y) that two randomly chosen alleles are identical is

$$j_y = \sum_{i=1}^{m} y_i^2$$

where $y_i = y_1, y_2, \ldots y_m$. The probability that two randomly chosen alleles, one from population X and the other from population Y, are identical is,

$$j_{xy} = x_1 y_1 + x_2 y_2 + \ldots + x_m y_m$$

or

$$j_{xy} = \sum_{i=1}^{m} x_i y_i$$

8

The genetic identity (I_j) of alleles between X and Y with respect to this locus is then defined as

$$I_j = \frac{j_{xy}}{\sqrt{j_x j_y}}$$

genetic identity, *I*

This quantity is unity when the two populations have the same alleles at identical frequencies, while it is zero when they have no alleles in common.

The identity of alleles over all loci studied is given by,

$$I = \frac{J_{xy}}{\sqrt{J_x J_y}}$$

where J_x, J_y and J_{xy} are the arithmetic means of j_x, j_y and j_{xy} respectively over all loci, including the ones with only one allele present.

The *genetic distance* (*D*) between X and Y is then defined as

$$D = -2.3 \log_{10} I$$

genetic distance , *D*

(Strictly, *D* is defined as the negative natural logarithm of *I*, i.e. $D = -\ln I$. These two formulae for *D* are exactly equivalent.) *D* is a measure of the average number of allele substitutions accumulated per locus.

EXAMPLE   Two populations X and Y differ in allele frequencies at two loci. At locus *A*, two alleles are present, A_1 and A_2; at locus *B*, three alleles are present, B_1, B_2 and B_3. The results of an electrophoretic survey (see Unit 10) are given in Table 1. Find the genetic distance between the populations. (Note: This example is given only as a simplified illustration of how genetic distances are calculated. Valid measures are based on data obtained from many more loci, to try to avoid sampling bias.)

TABLE 1   Allele frequencies in the *A* and *B* loci of populations X and Y

Allele	Allele frequency	
	population X	population Y
Locus *A*		
A_1	0.2	0.7
A_2	0.8	0.3
Locus *B*		
B_1	0.1	0.9
B_2	0.3	0.1
B_3	0.6	0

In order to calculate the genetic distance between X and Y with respect to the two loci, we proceed as follows.

First calculate j_x, j_y and j_{xy} for the two loci.

For locus *A*:

$$j_x = x_1^2 + x_2^2$$
$$= (0.2)^2 + (0.8)^2 = 0.68$$
$$j_y = y_1^2 + y_2^2$$
$$= (0.7)^2 + (0.3)^2 = 0.58$$
$$j_{xy} = x_1 y_1 + x_2 y_2$$
$$= (0.2 \times 0.7) + (0.8 \times 0.3) = 0.38$$

Similarly, for locus *B*:

$$j_x = x_1^2 + x_2^2 + x_3^2$$
$$= (0.1)^2 + (0.3)^2 + (0.6)^2 = 0.46$$
$$j_y = (0.9)^2 + (0.1)^2 + (0)^2 = 0.82$$
$$j_{xy} = (0.1 \times 0.9) + (0.3 \times 0.1) + (0.6 \times 0) = 0.12$$

The arithmetic means J_x, J_y and J_{xy}, over the two loci, are then obtained:

$$J_x = (0.68 + 0.46)/2 = 0.57$$
$$J_y = (0.58 + 0.82)/2 = 0.70$$
$$J_{xy} = (0.38 + 0.12)/2 = 0.25$$

The genetic identity, I, can then be obtained, as well as the genetic distance, D.

ITQ 2 Complete the calculations above in order to obtain an estimate of genetic distance.

One word of caution is required here. Although we refer to protein differences as 'genetic', they are in fact also a kind of 'phenotype'. Many molecular events interpose between the genes (DNA) and the product (proteins). Nevertheless, protein differences are clear and measurable. For this reason, they are regarded by some as more fundamental than differences in morphology which are often not easily quantifiable.

Some biochemical techniques and immunological reactions which also give estimates of I between populations are described in Section 4.2.3.

2.1.3 Reproductive isolation

Whether two populations can reproduce with one another is often considered the most important criterion for distinguishing between species. Isolating mechanisms are diverse, and there may be more than one mechanism operating between pairs of species. We shall treat these in detail in Section 2.2.

2.2 Species isolating mechanisms

What prevents genetic exchanges between species? There are many reproductive isolating mechanisms which can keep species apart. These are most conveniently classified as *pre-zygotic* (pre-mating) and *post-zygotic* (post-mating). The former prevent or impede hybridization, the latter reduce the viability or fertility of the hybrid.

The pre-zygotic mechanisms include:

pre-zygotic (pre-mating) isolating mechanisms

1 *Ecological or habitat isolation* Species occupy different habitats in the same geographic area.

2 *Seasonal or temporal isolation* Species in the same geographic area have different mating or flowering seasons, or become sexually mature at different times of the year.

3 *Ethological or sexual isolation* The sexual attraction between males and females of different animal species is reduced or absent due to mismatch in behaviour or physiology.

4 *'Mechanical' isolation* The mechanical structure of reproductive organs or genitalia impedes or prevents cross-fertilization between species.

5 *Gametic isolation* Either male and female gametes of different species fail to attract or unite with each other, or male gametes are inviable in the reproductive tracts of females of another animal species or on the styles of another species of plant.

The post-zygotic isolating mechanisms are:

post-zygotic (post-mating) isolating mechanisms

1 *Hybrid inviability* Zygotic development fails to take place, or zygotes become arrested at various stage of development, or the hybrid dies before reproduction.

2 *Hybrid sterility* The hybrid fails to produce functioning sex cells or gametes.

3 *Hybrid breakdown* The viability or fertility of the progenies of hybrids are reduced.

2.2.1 Examples of pre-zygotic isolating mechanisms

The separation of habitats plays an important part in the reproductive isolation of some species of the mosquito *Anopheles*. Several of the species in the genus are called *sibling species* because they are morphologically identical as adults and

sibling species

have in fact long been confused under the one name of *A. maculipennis*. They differ, however, in their prey preference (and hence ability to transmit human malaria) and in their habitat preference. *Anopheles labranchiae* and *A. atroparvus* live as larvae in brackish water, *A. maculipennis* in running freshwater, and *A. melanoon* and *A. messeae* in stagnant freshwater. The species differ also in their courtship rituals, which contributes to pre-zygotic isolation. Post-zygotic isolating mechanisms such as hybrid sterility are also found in crosses between some of the pairs of sibling species.

Seasonal isolation occurs between the toads *Bufo americanus* and *B. fowleri*; the former breeds earlier in spring than the latter, although there is some overlap and fertile hybrids are sometimes produced. These species are also isolated by their habitat preference; *B. americanus* lives in forested areas whereas *B. fowleri* lives in grasslands, so that in territories undisturbed by humans the species are well separated. However, the destruction of forests and the utilization of land for agriculture have led to more hybridization, and in some places hybrid populations are formed.

> **ITQ 3** What is a likely explanation for the increase in hybridization in disturbed areas?

Seasonal isolation often occurs among plants growing in the same geographic area. The pines *Pinus radiata* and *P. muricata* are found together on the Monterey Peninsula in California. The former sheds its pollen early in February, the latter in April. Hybrids are relatively rare and are less vigorous and fertile.

Ethological or sexual isolation occurs in many groups of animals, and is often the most powerful factor keeping apart closely related species in the same geographic area. Isolation is achieved in many insects and vertebrates through species–specific courtship displays—a combination of visual, vocal, and sometimes tactile and olfactory stimuli. An example of the latter are the pheromones of butterflies and moths which are powerful and specific chemical sex attractants. Behavioural isolation is illustrated for Hawaiian *Drosophila* species in the television programme 'The picture wings of Hawaii'.

'Mechanical' isolation frequently occurs among flowering plants. Because of the structure of their flowers, some can only be pollinated by a few or even a single species of insect. An extreme example of this is seen in the orchid family, where the richness and diversity of the species is believed to be due in part to their one-to-one relationship with specific insect pollinators. Among the more interesting insect pollinators of orchids are certain species of bees. The flowers attract sexually active males by their resemblance in colour and form to female bees of the same species; the males attempt to copulate with the flowers and, in so doing, pollinate them.

Gametic isolation is frequently found in marine organisms which shed their eggs and sperm into water; the normal attraction between eggs and sperms of the same species may be reduced or absent when different species are involved. Often, among flowering plants, the pollen of one species fails to germinate on the stigma of another species, or the pollen tube grows more slowly through the style of a different species, or it may start growing and then stop.

2.2.2 Examples of post-zygotic isolating mechanisms

Hybrid inviability may occur at any stage of development following fertilization. In crosses between sheep and goats, for example, the embryos die at an early developmental stage. In other crosses where the hybrids are viable, they may suffer from a reduction in fitness, particularly where sexual isolation between species is strong. Sterility of hybrids is frequently observed in interspecific crosses. The classic example is the mule, the progeny of a cross between a female horse (*Equus caballus*) and a male ass (*Equus onager*). Hybrid sterility is often partial and confined to one sex, usually the male; this is seen in many interspecific crosses of *Drosophila*. The failure of the progenies of hybrids to survive or to reproduce (*hybrid breakdown*) is an additional barrier to gene exchange between species. In crosses between *D. persimilis* and *D. pseudoobscura* the F_1 females are vigorous and fully fertile, yet when they are crossed to the males of either parental species the offspring are weak and often sterile. Hybrid breakdown is probably due to the formation of inferior genotypes by recombination between the genotypes of parental species.

hybrid inviability

hybrid sterility

hybrid breakdown

ITQ 4 Can you recall from the Science Foundation Course when it is that recombination between the parental genotypes occurs?

2.3 Difficulties with the species concept

It is by no means clear that the biological species concept itself is a useful taxonomic tool, because difficulties arise in trying to apply the criterion of reproductive isolation.

Plant taxonomists are by far the most frequent critics of the biological species concept. Serious doubts have been expressed as to the reality of plant species in nature. Some general objections to the concept are summarized below.

1 The existence of a species as an integrated gene pool, or as a population integrated by bonds of mating, is not substantiated by the evidence. It is rare for pollen to be dispersed from the parent plant for more than 1 000 m, and even when pollen grains are transported long distances the pollen is often no longer viable because pollen grains normally remain viable for only a few hours. This restriction of gene flow to immediate neighbourhoods also applies to many animal populations. However, this objection may not be serious because the movement of individual animals, pollen grains or propagules (e.g. seeds) over many generations may result in the spread of alleles through the species at a rate quite rapid enough to make the species a cohesive unit.

2 The concept of the species as a reproductively isolated unit frequently breaks down in plants. Many problems in the delimitation of species are created by hybridization in nature.

3 The lack of agreement between reproductive isolation and morphological or genetic *divergence* (as measured by protein differences, see Section 3.2.2) between species raises considerable doubt about the usefulness of the species concept. This is a more serious and fundamental objection than the others and we shall return to it in Sections 4 and 5.

2.4 Objectives and SAQs for Section 2

Now that you have completed this Section you should be able to:

(a) Define a species on biological criteria.

(b) Evaluate the different criteria for recognizing species.

(c) Recognize species isolating mechanisms.

(d) Recognize difficulties in applying the species concept.

To test your understanding of this Section, try the following SAQs.

SAQ 1 (*Objectives a, b and d*) For each of the cases (a)–(d) listed below, give reasons why the criterion of reproductive isolation fails to distinguish between species.

(a) Asexual species;

(b) palaeontological species;

(c) species with non-overlapping geographical distributions, i.e. allopatric species;

(d) species with overlapping geographic ranges which hybridize freely in the overlapping part of their distributions.

SAQ 2 (*Objective c*) Two species, *Drosophila silvestris* and *D. heteroneura*, are widely distributed with considerable overlap on the Big island of Hawaii. In nature, reproductive isolation is almost 100 per cent effective, but hybrids can be formed in the laboratory under 'no-choice' conditions using single interspecific mating pairs. The results of a mating experiment involving the two species and the hybrid progeny are summarized in Table 2. From the data in the Table, what is the mechanism isolating *D. silvestris* from *D. heteroneura*?

TABLE 2 Hybridization between *D. silvestris* and *D. heteroneura* from Hawaii, under laboratory conditions

Mating combinations*	Number of pair matings	Number of successful matings	Numbers of each sex in sample of adult progeny	
			males	females
heteroneura ♀ × *silvestris* ♂	36	0	—	—
silvestris ♀ × *heteroneura* ♂	28	8	80	78
F_1 ♀ × *heteroneura* ♂	10	5	40	38
F_1 ♀ × *silvestris* ♂	10	1	—	—
heteroneura ♀ × F_1 ♂	10	5	39	37
silvestris ♀ × F_1 ♂	10	5	43	33
F_1 ♀ × F_1 ♂	9	5	54	42

* ♀ is female; ♂ is male.

3 The population structure of species

Study comment This Section describes the patterns of intraspecific variation at the morphological, physiological and genetic levels, and investigates the forces responsible for producing these patterns. You will need graph paper for work in Section 3.2.2.

A species extends and varies over both space and time. Although we tend to regard only changes over a period of time as evolution, the variation of a species over its range of distribution is itself an evolutionary phenomenon. In this Section we shall examine the extent and nature of geographical variation within species, in the hope that this will yield important clues as to the nature and causes of evolutionary change.

3.1 Intraspecific variation

3.1.1 Patterns of variation

The variation that exists within a single population is but a fraction of that which is present in a species over its whole range. Typically, a species is divisible into a number of populations showing intergradations of morphological or physiological characters (i.e. 'clinal variation', see Section 3.1.2) in its main range. Some geographically isolated populations may be present here and there (see Figure 1). Within the main range, collections of populations may show greater or lesser discontinuities with contiguous populations in some physiological or morphological traits, and there are often narrow zones where hybrids occur between such contiguous groups.

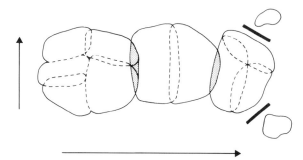

FIGURE 1 Population structure of an idealized species. Each area marked off by a dotted line represents a population or a microgeographic race. Each area marked off by an unbroken line constitutes a subspecies or semispecies. Grey areas represent hybrid zones. Thick solid lines represent geographic barriers. Arrows represent the direction of clinal variations.

13

Populations or groups of populations that appear more or less distinct constitute various subdivisions within the species. A *microgeographic race* is the lowest subdivision, and may correspond to a single population. The size of a microgeographic race will depend on

(i) the mobility of the organisms (in animals), or the distances over which pollen or propagules are transported (in plants);

(ii) the size of the microgeographic environments to which a population may be adapted; and

(iii) the factors preventing or reducing gene flow (exchange of genes) between neighbouring populations.

Microgeographic races are often produced by adaptation to extremely localized environmental conditions. For example, races of the grasses *Agrostis*, *Festuca* and *Anthoxanthum* that are tolerant to heavy metals have developed on mine wastes in which there are high concentrations of copper, zinc or lead. The areas where they occur are seldom more than 500 m across. The boundaries between tolerant and non-tolerant (normal) varieties are often remarkably sharp (corresponding to the exact boundaries between the contaminated and non-contaminated areas), despite wind pollination which encourages gene flow across the boundaries.

The next level of subdivision within a species is the *macrogeographic race* or *subspecies*. These terms apply to larger conglomerates of populations over wide geographic areas. Some taxonomists (chiefly zoological) have agreed that a species should be divided into subspecies, which should be given taxonomic names, if a majority (say 75 per cent) of individuals in the species can be recognized as belonging to one or other of the subspecies. According to estimates published in 1970 there are approximately 8 600 species of birds divided into about 28 500 subspecies, or 3.3 subspecies per species on average. Of course, some species are *monotypic* (having only one type) and consist of a single more or less distinct form; while others are highly *polytypic* (having more than one type) and consist of many distinguishable forms. Another subdivision is that of *semispecies*— morphologically distinct groups that are partially reproductively isolated in nature. There is a tendency to regard the various subdivisions within a species as stages in the formation of new species.

3.1.2 Clinal variation

Superimposed over the population structure of species are clinal variations (see Figure 1), that is, variations that exhibit a gradient or gradation across the range of the species. For example, when neighbouring populations of a species are compared, they usually differ from each other in a number of characteristics, and when certain individual characters are traced through a series of adjacent populations, the changes usually show a regular progression. Julian Huxley coined the term *cline* for such a character gradient. Clines also occur vertically; for example, altitudinal clines extend from lowlands to mountains, and in the ocean they extend through different depths.

One of the best analyses of clinal variation in animals is Peterson's study of Fenniscandian butterflies. Of 59 characters (in 16 species of butterflies) 29 varied clinally. The six migratory species exhibited no clines, but 70 per cent of the characters examined in non-migratory species varied clinally. Another well known cline exists in Australian birds, which decrease in size from Tasmania northwards to the Torres Straits (i.e. towards the Equator). This is an example of *Bergmann's rule*, which states that in geographically variable species of warm-blooded animals the body size is larger in the cooler parts of the range of the species. Clines also exist for physiological characters. For example, R. B. Goldschmidt found that in the Northern Hemisphere the duration of larval development in the gypsy moth, *Lymantria monacha*, exhibits a north–south cline, being longer in the south. This was correlated with the length of the period of optimum temperature for development, i.e. the summer period.

The existence of clines does not always correlate with environmental gradients, and even where a correlation is found it is not always easy to prove that it represents an adaptation of the species to that environmental gradient. One factor which contributes to the formation of clines is gene flow, or the slow diffusion of genes across population boundaries due to the dispersal of gametes or of zygotes. Thus, a population in which a certain allele exists at a high frequency could produce a gradient for that allele through a series of neighbouring populations in which the allele was originally absent or present at very low frequencies. Such clines are frequently formed when migration between populations has taken place. A well known example is the cline in frequency of human B blood-group alleles, which decreases from east to west in Asia and Europe. It is likely that this cline is the result of migration.

In summary, clines do not automatically imply an adaptation to an environmental gradient. It is often impossible to distinguish between an adaptive gradient which implies an equilibrium pattern, and a historical transient pattern due to migration or to spread of genes from their point of origin.

3.1.3 Geographic isolates

Geographic isolates are populations or groups of populations prevented by a geographic barrier from free gene exchange with other populations of the species. The isolation is rarely complete, though the amount of gene exchange is extremely limited compared with that between populations in the main range.

Virtually all species contain some isolates, particularly near the periphery of the species range (peripheral isolates). The frequency of isolates increases sharply whenever geographic or ecological conditions favour an insular distribution pattern. This is true not only for oceanic islands but also for ecological 'islands' such as mountain tops, forest patches in grassland, or lakes and streams. Species often show great phenotypic uniformity over wide areas where the species range is continuous, and an astonishing production of distinctively different isolates where barriers break up the range. The importance of isolation (whether in peripheral isolates or in geographic or ecological islands) in producing genetic change is attributed to the following:

peripheral isolates

1 divergence resulting from founder effects or random genetic drift (see Unit 10);

2 rapid changes in the colonizing populations due to temporary release from competition or from factors which limit their population size (Unit 11);

3 adaptation to the local geographic and climatic conditions;

4 adaptation to the local biotic conditions (the existing flora and fauna);

5 cessation of gene flow.

The various factors favouring genetic change in isolates are of great importance in the process of speciation, as we shall see in Section 5.

3.1.4 Hybrid zones

Hybrid zones are areas of contact between phenotypically different populations where there is some gene flow between the populations. Taxonomists have sometimes referred to these belts as subspecies borders. Consider the distribution of the carrion crow (*Corvus corone corone*) and the hooded crow (*C. c. cornix*), both considered subspecies of *C. corone*. The all-black carrion crow inhabits western Europe including Scotland and eastern Ireland; the hooded crow, grey with a black head, wings and tail, inhabits central and eastern Europe, northern Scotland and western Ireland. Where the two subspecies meet there is a zone of hybridization varying from 24 to 170 km in width (see Figure 2). Pairing within the

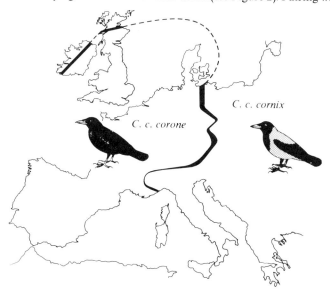

C. c. cornix

C. c. corone

FIGURE 2 Position of the hybrid zone between the carrion crow (*Corvus corone corone*) and the hooded crow (*Corvus corone cornix*) in western Europe. Note the relative narrowness and unequal width of the zone.

hybrid zone appears to be random, and every conceivable combination of parental characters and all degrees of hybridity are present. Hybrids are rarely found outside the hybrid zone.

Many cases of secondary hybridization occurred when geographically isolated populations produced as a result of glaciation in the Pleistocene expanded and came into contact in recent times. At the height of glaciation, the ranges of many temperate zone species contracted into small pockets called *glacial refuges*. In Europe, for instance, the Alpine and northern icecaps came within 500 km of each other, separated by icy windswept steppes, and the forest animals retreated into refuges in southwestern or southeastern Europe. When conditions improved at the end of the Pleistocene and the populations in the refuges expanded northwards, the isolates had diverged genetically and had become sufficiently distinct for hybrid zones to form along the line of contact between the two populations. Such hybrid zones in Central Europe have been identified for mammals, birds, amphibians and invertebrates.

glacial refuge

3.2 The polytypic species

We shall examine in detail two examples of intraspecific variation, in order to focus on some of the physical and biotic factors involved in producing the existing patterns of variation.

3.2.1 Mimicry in *Heliconius*

The South American butterflies of the genus *Heliconius* have been studied for more than fifteen years by J. R. G. Turner and his colleagues. The genus *Heliconius* contains a number of species, each with a series of macrogeographic races. The races differ markedly from one another in the colour patterns of their wings. What is most striking is the parallel variation in wing pattern occurring in races of the two different species, *Heliconius erato* and *H. melpomene* (see Figure 3). At any one locality in Latin America the wing patterns of the races from the two different species show great similarities to each other. As both species are distasteful to birds, there is a strong possibility that this parallel series represents an extended system of *Mullerian mimicry*—the mutual mimicry (resemblance) of unpalatable species which probably affords them increased protection from predators. As both species are distasteful and have warning coloration, they may be subject to stabilizing selection (see Unit 10) through predation.

Mullerian mimicry

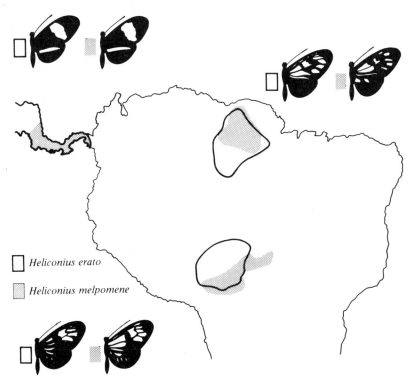

FIGURE 3 Parallel variation in three races of *Heliconius melpomene* (grey areas) and *Heliconius erato* (areas enclosed by heavy black lines). Very similar wing patterns for the two species are found in the same general areas.

16

□ What would this hypothesis suggest concerning the fate of bad mimics?

■ If stabilizing selection exists, butterflies which deviate from the usual pattern and are therefore bad mimics are more likely to be caught and eaten than normal individuals. In other words, there will be selection against them.

To test this, W. W. Benson applied paint to the wings of *Heliconius erato* in the rain forest in Costa Rica in order to alter their wing patterns. Controls had the same amount of paint applied without altering their wing patterns. The butterflies were then released. Samples were later recaptured to assess the extent of predation by birds. The results are shown in Table 3.

TABLE 3 Wing damage in altered and unaltered (control) individuals of *H. erato* on recapture

	Males		Females		Both sexes	
	altered	unaltered	altered	unaltered	altered	unaltered
Total no. of butterflies recaptured	12	14	4	3	16	17
No. of damaged butterflies recaptured	4	0	2	0	6	0

ITQ 5 How would you interpret these results?

In both species, each race presents an almost uniform wing pattern throughout its range of distribution. Where races of one species meet, hybrid zones are formed. The hybrid zones are extremely narrow, perhaps because stabilizing selection for the parental patterns leads to the elimination of the first generation hybrids and subsequent recombinants. Such apparent stabilizing selection is often augmented by natural ecological barriers to hybridization.

The obvious explanation of the formation of races is that each evolved in isolation, which could have occurred after glaciation restricted a once widespread population to small forest 'refuges'. But how does isolation in itself produce such a divergence of wing patterns in the different races of both species? In order to answer this question, it is necessary to know something about the genetics of colour patterns and how Mullerian mimicry may have evolved.

Two theories have been proposed for the evolution of Mullerian mimicry. One, by Dixey, maintains that the patterns of two warningly coloured species evolve gradually towards one another until they both take up some intermediate pattern. The other theory, put forward by Marshall, suggests that the less protected but unpalatable species takes over the pattern of the better protected species which would never evolve towards that of the less protected species. In terms of genetic changes, Dixey's theory would involve a gradual convergence to a pattern by means of an accumulation of *polygenes*, or many genes each with a small effect on the pattern; the changes required by Marshall's theory would be accomplished by the substitution of a few genes, each with large effects in changing one wing pattern to another. In other words, the choice is between a *gradual*, more or less symmetrical convergence of the two species to an intermediate wing pattern, and a saltatory (sudden) asymmetrical conversion of the pattern in one species to that of the other, better protected species.

polygenes

Inter-racial crosses in both *Heliconius* species have shown that wing colour-pattern is controlled by a small number of genes with large effects (called major genes), with a further refinement of the pattern produced by 'modifying' polygenes.

□ Which of the hypotheses on the evolution of mimicry is supported by this genetic evidence?

■ The genetic data are more consistent with an asymmetric conversion of one pattern to another through the acquisition of a few major genes. However, it is not known which of the species in each geographic area was originally the better protected model, towards which the other evolved.

It is probable that one of the most important causes of divergence in *Heliconius* is adaptation to the local biotic environment, more specifically to the presence of another, better protected, species. R. MacArthur and E. O. Wilson point out that islands will differ from one another and from the mainland because the composition of the flora and fauna is subject to a continual and more or less random process of extinction and recolonization (see Unit 11). It is therefore reasonable to assume that the rain forest refuges of Amazonia, being true ecological islands, will by chance differ in their fauna and flora. The main factor responsible for this must have been differential extinction in the islands, because the original forest would have had the full complement of plants and animals.

Race formation as it occurs in *Heliconius* is very likely to be a prelude to the formation of new species. There are a number of other species of butterflies in South America with races analogous to one or more of the parallel races in *melpomene* and *erato*. This is to be expected if races become full species by developing isolating mechanisms.

3.2.2 Genetic differentiation in *Drosophila* species

Geographical races are usually morphologically distinct from one another. But the question which is uppermost in the minds of many evolutionists is: How much genetic difference exists between populations, and what is their level of genetic divergence?

The most comprehensive study in this area was carried out first by T. Dobzhansky then by F. J. Ayala and his co-workers on the *Drosophila willistoni* complex of populations and related species. The *willistoni* group consists of more than a dozen closely related species native to the American tropics. Some species of the group, such as *D. nebuloso*, are morphologically distinct from the rest. Others are called sibling species because they are morphologically very similar, even though they are completely reproductively isolated. Two sibling species, *D. insularis* and *D. pavlovskiana*, are narrow endemics, the former to some islands of the Lesser Antilles, the latter to Guyana. Four other species, namely *D. equinoxialis*, *D. tropicalis*, *D. willistoni* and *D. paulistorum*, are siblings and have widely overlapping geographic distributions through Central America, the Caribbean, and much of South America.

Some species consist of more than one subspecies. *Drosophila willistoni willistoni* and *D. w. quechua* are separated by the Andes in Peru. Hybrid males but not hybrid females of laboratory crosses are sterile when the female parent is *D. w. quechua*, but there is no evidence of behavioural isolation between the subspecies. Another pair of subspecies, *D. equinoxialis caribbensis* and *D. e. equinoxialis*, show no behavioural isolation but always yield sterile males and fertile female hybrids in breeding experiments. These subspecies show some degree of post-zygotic isolation and could be considered to be incipient species.

Another set of populations, possibly at a more advanced stage in the speciation process, is represented by six semispecies of *Drosophila paulistorum*. In many localities two or more semispecies coexist sympatrically, and yet reproductive isolation is almost complete. In addition to the partial sterility of hybrids produced in laboratory crosses, sexual isolation through behavioural differences is also well developed between some semispecies.

Electrophoretic variation (see Unit 10) in proteins encoded by 36 gene loci were studied in populations of several species of the *willistoni* group.

> **ITQ 6** (a) What is the basis for regarding electrophoretic variation in proteins as an estimate of genetic variation?
>
> (b) What are the limitations of this method? Try to recall three.

From the differences in allele frequencies at these loci, estimates were made of genetic distances between populations at each level of evolutionary divergence. The results are given in Table 4. From these data, there appears to be a more or less distinct gradation in the level of genetic divergence corresponding to each level of evolutionary divergence.

TABLE 4 Average genetic identity, $\bar{I}$, and average distance, $\bar{D}$, between taxa having various levels of evolutionary divergence in the *Drosophila willistoni* group

Taxonomic level	Genetic identity, $\bar{I}$	Genetic distance, $\bar{D}$
local populations	0.970 ± 0.006	0.031 ± 0.007
subspecies	0.795 ± 0.013	0.230 ± 0.016
semispecies*	0.798 ± 0.026	0.226 ± 0.033
sibling species*	0.563 ± 0.023	0.581 ± 0.039
non-sibling species	0.352 ± 0.023	1.056 ± 0.068

* These $\bar{I}$ and $\bar{D}$ values are based only on well sampled semispecies or sibling species and exclude data from *D. insularis* and *D. pavlovskiana*.

TABLE 5   Genetic distances between some species of the *Drosophila willistoni* group*

	D. willistoni	D. tropicalis	D. equinoxialis	D. paulistorum	D. pavlovskiana	D. insularis
D. willistoni	—					
D. tropicalis	0.413	—				
D. equinoxialis	0.656	0.665	—			
D. paulistorum	0.524	0.609	0.621	—		
D. pavlovskiana	0.518	0.701	0.633	0.232	—	
D. insularis	1.070	0.883	1.091	1.208	1.273	—

* When subspecies or semispecies occur, the average values between each one of them and the other species are given. The first four, and th last two, of the species listed are sets of sibling species.

Genetic distances can be used to construct *dendrograms* (branching tree diagrams) representing phylogenetic relationships among the groups. In constructing dendrograms, a number of different procedures (called cluster analyses) are used, one of which is illustrated by the following example. Table 5 gives some data on genetic distances between sibling species of the *Drosophila willistoni* complex. Genetic distances between pairs vary from 0.232 to 1.273. On a piece of graph paper, mark out on the horizontal axis the appropriate scale of genetic distance from 0 to 1.3, *right to left*. Note that the least genetic distance separates the species pair, *D. pavlovskiana* and *D. paulistorum*: draw a short vertical line at the genetic distance of 0.232 and draw two prongs horizontally from the ends of the vertical line towards the edge of the paper on the right. Label each prong with the name of one of the two species concerned. Leave room to fill in the other species, see Figure 4(a).

dendrograms

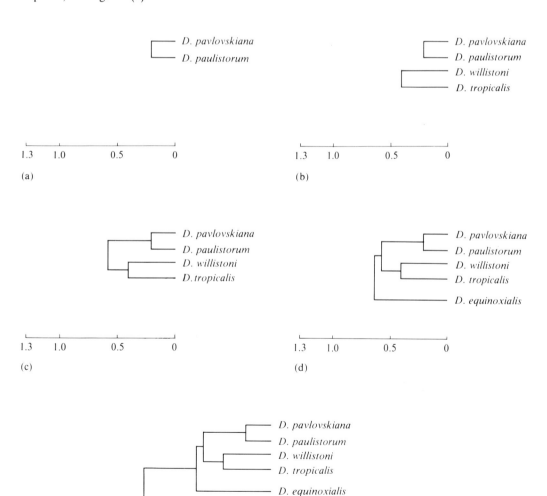

FIGURE 4   Constructing a dendrogram from genetic distances. For details, see text.

The next smallest genetic distance (0.413) is between *D. willistoni* and *D. tropicalis*. On the graph, make a small vertical line at genetic distance 0.413 and draw horizontal lines for the two species as before (Figure 4b). You now have four species in two pairs and we want to compute how far apart, *on the average*, the two pairs are from each other. To do this you need to find the average of the four distances separating *D. willistoni–D. tropicalis* from *D. paulistorum–D. pavlovskiana*. Thus *D. willistoni* is separated from *D. paulistorum* by 0.524, and from *D. pavlovskiana* by 0.518: whereas *D. tropicalis* is separated from *D. paulistorum* by 0.609 and from *D. pavlovskiana* by 0.701. The average of these four values comes to 0.588. At a genetic distance of 0.588, make a vertical line which connects up to the two species pairs (see Figure 4c).

The next species to join up is *D. equinoxialis*, as its genetic distance from each of the four species already on the graph is considerably less than that of the remaining species, *D. insularis*. Again you need to find the average distance of *D. equinoxialis* from the four species. These four values are: *D. equinoxialis–D. willistoni*, 0.656; *D. equinoxialis–D. tropicalis*, 0.665; *D. equinoxialis–D. paulistorum*, 0.621; *D. equinoxialis–D. pavlovskiana*, 0.633. The average is 0.644. It can now be joined up as in Figure 4(d). The last species, *D. insularis*, has five distances from the others, namely 1.070, 0.883, 1.091, 1.208 and 1.273, giving an average of 1.105. The dendrogram can be completed as in Figure 4(e).

As you may have noticed, the dendrogram distorts the true values of the genetic distances between species considerably because of the averaging process. This means that species represented at a certain distance from a cluster on the dendrogram may be more closely related than is apparent to some of the species within the cluster. Thus *D. equinoxialis*, which is outside the cluster of the four species *D. paulistorum*, *D. pavlovskiana*, *D. willistoni*, and *D. tropicalis*, is in fact more closely related genetically to each one of them than *D. tropicalis* is related to *D. pavlovskiana*. You may also have noticed that, on this dendrogram, some sibling species are more widely separated than others.

> **ITQ 7** In the dendrogram, which sibling species are more widely separated from each other than are species which are non-sibling?

One conclusion which can be drawn from this study is that morphological divergence between populations is poorly correlated with genetic divergence as measured by differences in proteins. Thus sibling species (from morphological considerations) may be genetically more distant than non-sibling species. Indeed, subsequent studies involving other groups of *Drosophila* and invertebrates show that there is generally a wide range of genetic divergence existing at each level of evolutionary divergence, and there is a tendency for some ranges to overlap considerably.

3.3 Objectives and SAQs for Section 3

Now that you have completed this Section, you should be able to:

(a) Describe examples of typical patterns of variation in *Heliconius*.

(b) Evaluate the significance of physical and biotic factors in producing evolutionary changes in *Heliconius*.

(c) Construct and interpret dendrograms from data on genetic distance.

To test your understanding of this Section, try the following SAQs.

SAQ 3 (*Objectives a and b*) Which of the following factors were probably the most significant in the evolution of the parallel races of *H. erato* and *H. melpomene*?

  (i) Founder effect or random genetic drift in small isolated populations;

  (ii) temporary release from competition in isolation;

  (iii) adaptation to local physiographic and climatic conditions;

  (iv) adaptation to local biotic conditions;

  (v) cessation of gene flow.

SAQ 4 (*Objective c*) Use the data in Table 6 to construct a dendrogram from the genetic distances measured for six populations of *Drosophila subobscura* obtained in a survey of the British Isles.

TABLE 6   Genetic distances of six populations of *Drosophila subobscura*

	A	B	C	D	E	F
A	—					
B	0.220	—				
C	0.299	0.300	—			
D	0.600	0.550	0.498	—		
E	0.550	0.499	0.600	0.300	—	
F	0.480	0.450	0.500	0.350	0.140	—

4   Interspecific variation

Study comment   This Section examines interspecific differences at the chromosome, gene and whole organism levels and the relationships between them. The main point to note is that, although differences exist at all levels, there is no simple relationship between any two levels. This Section is relatively demanding and requires more effort and concentration than any other Section. You should study AV 8, 'Reconstructing picture wing evolution', after reading Section 4.1.1.

Species differ in morphology and physiology, and in their ecological requirements. It is natural to assume that such differences reflect those existing at the genetic level. Attempts to define genetic differences between species were for a long time frustrated by a lack of suitable tools. By definition, most species do not hybridize in nature, and although some do hybridize in the laboratory the hybrids are often wholly or partially sterile. Thus traditional genetic analysis, involving crosses between individuals from different species which differ in one or more characters, cannot be satisfactorily performed. Even when crosses have been made, the results can rarely be interpreted simply. Often, the characters under study are not controlled by the same genes in the two species. This genetic gap between species so impressed Goldschmidt that he rejected the Darwinian picture of gradual evolution in favour of speciation by '*systemic mutations*', i.e. mutations giving rise to large phenotypic effects. Such mutants he called 'hopeful monsters'.

systemic mutations
'hopeful monsters'

> **ITQ 8**   Most biologists find Goldschmidt's 'hopeful monsters' very implausible. Can you think why?

Although Goldschmidt's solution to the problem could easily be dismissed, his arguments showing that a problem does exist are by and large sound. In particular, he emphasized two important aspects of evolution which the theory of natural selection does not adequately explain: the origin of *reproductive gaps*, and the origin of *morphological discontinuities*. These two are not always related. We shall say more about the origin of morphological discontinuities in Unit 13. Here we will concentrate on the nature of reproductive gaps between species.

reproductive gaps
morphological discontinuities

The reproductive gaps to which Goldschmidt referred are the various *post-zygotic isolating mechanisms* between species (see Sections 2.2 and 2.2.2). Some of these mechanisms can be attributed to chromosomal incompatibility which causes a breakdown in meiosis. Others have been attributed to a general incompatibility of the different genomes. Ernst Mayr, for example, argued that genes do not function in isolation but as a 'co-adapted complex': because each species has a long evolutionary history in which its genes have been selected mainly for harmonious interactions with one another in development, the mixture of genes in a hybrid genome does not function as a unified whole.

co-adaptation of genes

The co-adaptation of the genome implies a functional interdependence between genes (a subject discussed further in Unit 13). Here we shall concentrate on the two main areas of research into interspecific differences that have emerged in recent years—chromosomal and molecular genetic studies—and on the evolutionary implications of these findings.

4.1 Karyotype differences between species

(a) The karyotype

The number and appearance of chromosomes at metaphase of mitosis in a eukaryote cell is referred to as its *karyotype*. In multicellular organisms all body or somatic cells, i.e. all the cells except gametes, have the same karyotype and most individuals within a species have the same somatic karyotype. Exceptions occur, for example where chromosomal polymorphism is present and where males and females have different sex chromosomes.

The characteristic features of a karyotype are:

(i) *The number of chromosomes*   It is customary to represent this number as $2n$ for diploid organisms, where n is the haploid number (see Section 2.3 of Unit 7). For the human karyotype, $2n = 46$ (Figure 5).

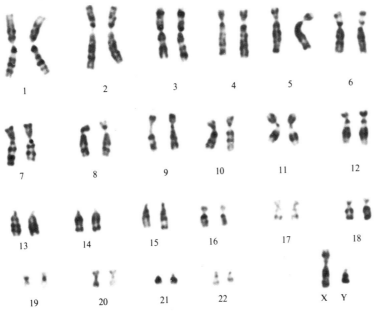

FIGURE 5 Photomicrograph of the karyotype of a human male stained with quinacrine to reveal the banding pattern.

(ii) *Chromosomal morphology*   Besides the actual shape and size of the chromosome, one of the most useful markers is the position of the centromere, a structure which appears as a constriction in the chromosome. Chromosomes with the centromere near the middle are *metacentric*, those with the centromere near one end are *acrocentric*, and those with a terminal centromere are *teleocentric*.

chromosomal morphology

(b) Chromosomal mutations

Chromosomal mutations involve changes in chromosome structure or number. Two common structural mutations, *inversions* and *translocations*, and their effects on meiosis are summarized in Figures 6 and 7. An inversion (Figure 6) is a rearrangement within a single chromosome in which a segment has been rotated through 180°. A translocation (Figure 7) is a transfer of segments usually by reciprocal exchange between different chromosomes. Both inversions and translocations can lead to the formation of non-viable gametes.

chromosomal mutation

inversion

translocation

Fission of chromosomes occurs when a two-armed (metacentric) chromosome is converted to two one-armed (teleocentric) chromosomes; conversely, *fusion of chromosomes* takes place when two teleocentric chromosomes join together to give a metacentric one. The occurrence of both fission and fusion (and often it is unclear which of these is the primitive condition) makes it necessary when describing karyotypes to cite the number of major chromosome *arms*.

fission of chromosomes
fusion of chromosomes

The importance of such structural changes in evolution is that it often facilitates the reproductive isolation of populations. Thus, in an individual heterozygous for a large inversion in which crossing over occurs, meiosis is irregular and about half the gametes are defective (Figure 6). This reduces fertility in the heterozygote; the parental chromosomes and therefore gene combinations remain more or less intact. When the inversion involves only a short segment of the chromosome, (i) the chances of crossovers *within* the inversion are much reduced so that a high proportion of normal gametes may be produced, and (ii) the group of genes within the inversion tends to be conserved as a *supergene*.

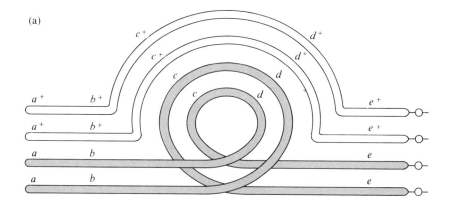

(a)

FIGURE 6 The behaviour of homologous
chromosomes in an individual heterozygous
for an inversion. (a) Pairing during meiosis.
(b) Crossing over. (c) Separation of
chromosomes at anaphase I, showing (i) a
chromatid bridge, (ii) a fragment without a
centromere (both (i) and (ii) give inviable
gametes), and (iii) two chromatids with all
parts intact (which give viable gametes).
Thus, half the gametes on average will be
defective in the heterozygote.

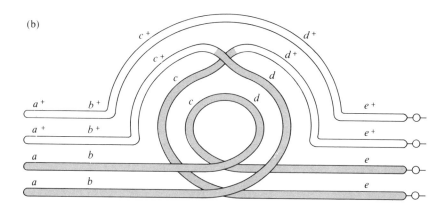

(b)

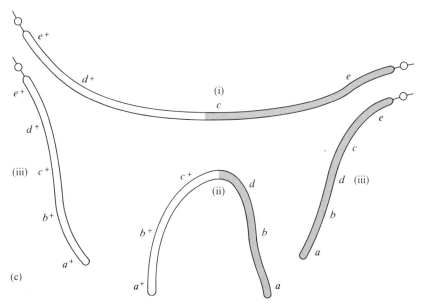

(c)

Polyploidy involves the multiplication of whole sets of chromosomes (each set being the haploid number n). Although not unknown in animals, polyploidy has much greater evolutionary importance in plants. Two distinct types of polyploid occur, autopolyploids and allopolyploids. *Autopolyploids* are polyploids in which there has been a straightforward doubling of the chromosome number. They are known to occur in nature and can readily be produced in laboratory conditions. A plant whose diploid genome is AA (having been formed from two gametes each containing the haploid chromosome complement A) may produce autopolyploid cells ($AAAA$) during development if, after duplication of its chromosomes prior to mitosis, cell division does not occur. Vegetative reproduction involving such cells gives rise to a wholly autopolyploid individual. Alternatively, if after meiosis there is a similar failure of cell division, gametes of genome AA are formed. These may meet and fuse with similar unreduced gametes, producing polyploid zygotes in which cell division and development can occur more or less as normal.

polyploidy

autopolyploids

23

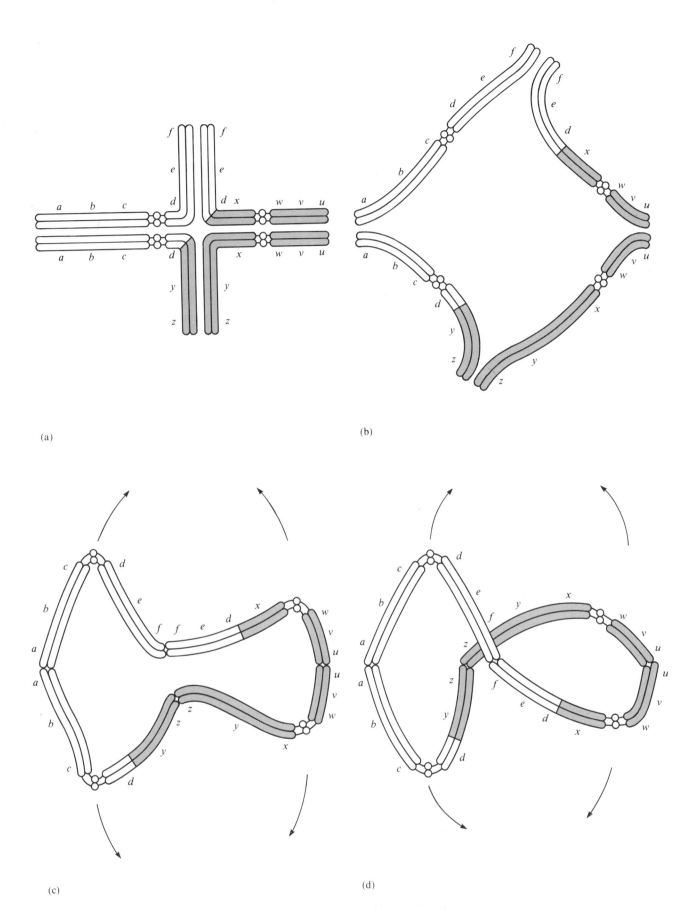

(a)

(b)

(c)

(d)

FIGURE 7 The behaviour of four chromosomes in an individual with a reciprocal translocation. (a) Pairing during meiosis. (b) Separation of the centromeres. (c) and (d) Alternative configurations at anaphase I: (c) gives rise to inviable gametes because one chromosome segment is duplicated and another is missing in both sets of segregating chromosomes; (d) gives rise to viable gametes as both upper and lower combinations are balanced, i.e. they contain all parts of the two chromosomes. Thus, assuming (c) and (d) to be equally likely configurations, on average half the gametes produced will be defective.

Allopolyploids are polyploids in which hybridization has been followed by an increase in the number of chromosomes. The hybrid is formed from two gametes containing dissimilar haploid genomes *A* and *B*, so the diploid hybrid is genetically *AB*. If now the genome is doubled the allopolyploid *AABB* will be produced.

allopolyploids

While naturally occurring *AAAA* autopolyploids are not rare in nature, very few of them are fertile. This is due to complications in meiosis when, instead of each chromosome pairing with *one* homologous partner, because the chromosome number has doubled, each chromosome now has three homologous partners with which it may pair. Consequently pairing may start between any two of the four homologues; this is usually followed by a failure of chromosomal separation, so normal fertile gametes are rarely produced.

Allopolyploids, on the other hand, may be fertile and reproductively isolated from both their parents. Here each chromosome has only one homologue with which to pair, so normal fertile gametes may be formed; but backcrosses to either parent are usually unsuccessful.

Unreduced gametes from allo- or autopolyploids may fuse with reduced gametes, producing zygotes with three sets of chromosomes. Plants developing from such zygotes are called either allo- or autotriploids. Fusion of two diploid gametes would give rise to an allo- or autotetraploid, with four sets of chromosomes ($4n$). In sexually reproducing species, only the even-numbered polyploids with at least two sets of chromosomes originating from any single species will be fertile. Odd-numbered types will be sterile due to irregularities in chromosome pairing and in assortment during meiosis.

Aneuploidy, the loss or gain of single chromosomes, occurs when paired homologues fail to separate during meiosis.

aneuploidy

4.1.1 Chromosomal inversions in *Drosophila* species

Genetic relationships between species can sometimes be inferred from chromosomal inversions, particularly if the inversions overlap. In *Drosophila* and other flies (Diptera), inversion and other chromosomal rearrangements can be studied with relative ease because the salivary gland cells contain giant chromosomes which have recognizable patterns of bands. These *polytene chromosomes* are described in more detail in AV 8, 'Reconstructing picture wing evolution'.

polytene chromosomes

☐ The homologous chromosomes of three different species show the following banding patterns:
Species 1   ABCDEFGH
Species 2   ABFGCDEH
Species 3   ABFEDCGH
Each band is represented by a letter, and the order of the letters represents the sequence of the bands. What are the possible evolutionary relationships between these species as indicated by these sequences?

■ The sequence of banding of species 1 could be changed to that of species 3 by a single inversion: CDEF → FEDC. Or the pattern of species 3 could be changed to that of species 1 by the reverse inversion: FEDC → CDEF. Similarly, the banding pattern of species 3 could be changed to that of species 2 by one inversion: EDCG → GCDE. The reverse inversion, GCDE → EDCG, would change the banding pattern of species 2 into that of species 3. But the banding pattern of species 1 cannot be changed to that of species 2 by means of an inversion, nor can the pattern of species 2 be changed to that of species 1. So the possible evolutionary relationships are:

$$1 \rightarrow 3 \rightarrow 2 \quad \text{or} \quad 2 \rightarrow 3 \rightarrow 1$$

In the absence of any other information, there is no way of telling which of these is the ancestral sequence.

Each chromosomal rearrangement in *Drosophila* may be regarded as having a single origin because of the following considerations:

(i)   There are about 5 000 bands in the polytene chromosomes.

(ii)   In order to have an inversion, two breaks must occur.

(iii) The spontaneous rate of inversion is one in 500 per individual.

25

☐ (a) How many possible inversions can occur in *Drosophila*?

(b) What is the probability that a particular inversion will arise?

■ (a) Each break can occur anywhere between each of the (approximately) 5 000 bands. As there have to be two breaks, the number of possible double breaks, and therefore possible inversions, is about 5 000 × 5 000 or 25 000 000.

(b) Since the spontaneous rate of inversions is one in 500, the probability that a particular rearrangement arises is $(1/500) \times (1/25\,000\,000)$, or $1/1.25 \times 10^{-10}$.

This means that 1.25×10^{10} individuals will have to be bred in order to have a reasonable chance of getting a second occurrence of the same rearrangement, so that each rearrangement is essentially unique. For this reason, inversions are invaluable in determining phylogenetic relationships between populations and species.

Now study AV 8, 'Reconstructing picture wing evolution', where you will see how chromosome mutations are used to make inferences about the phylogenies of some Hawaiian drosophilids.

4.1.2 Karyotype evolution in higher plants

Polyploidy is generally accepted to be important in speciation in higher plants, and will be dealt with in Section 6. Aneuploidy, inversions and translocations also occur frequently in higher plants. One example of special interest is provided by the extensive work on American species of the genus *Clarkia* (family: Onagraceae) by H. Lewis and his co-workers.

Clarkia

Though a few species of *Clarkia* appear to have been established in certain areas for a long time, evidence from two decades of study indicates that speciation in *Clarkia* is in many cases a recent phenomenon. The new species are usually distributed in a small area adjacent to their respective parent species. Morphologically, they resemble their parent species but are separated from them by strong reproductive barriers. In all instances, hybrids between the new species and their parents have much reduced fertility as a consequence of meiotic irregularities, indicative of chromosome reorganizations such as translocations and inversions. Each new species characteristically occupies a habitat that is drier than or ecologically marginal to that of the parent species.

We shall examine the case of *Clarkia biloba* and *C. lingulata* in detail. The former is a polytypic species with a wide area of distribution; the latter is narrowly confined to two colonies each of several thousand individuals in an area at the southern limit of distribution of *C. biloba*. Both species are insect pollinated. The anthers mature before the stigma, so that outcrossing is the norm, though individuals of either species are self-fertile and self-pollination can occur. The seeds have no special dispersal mechanism and are generally dropped close to the parent plant.

Clarkia biloba consists of three geographical races or subspecies, *brandegeae*, *biloba* and *australis*, which differ from one another in petal shape and colour. Of the three subspecies, *australis* most closely resembles *C. lingulata*. *Clarkia lingulata* has been crossed to each of the subspecies of *C. biloba*. F_1 hybrids were obtained most readily with *australis* and least readily with *brandegeae*. However, the fertility of the F_1 hybrid was greater when crossed with *brandegeae* than with either *australis* or *biloba*. Morphologically, the F_1 hybrids are intermediate in all respects.

Meiosis in *Clarkia biloba*, *C. lingulata* and their hybrids has been examined. All the subspecies of *C. biloba* have 8 pairs of chromosomes, whereas *C. lingulata* possesses 9 pairs (Figures 8a and b). Hybrids between *C. lingulata* and the three subspecies show irregularities in the configuration of chromosomes at metaphase, a stage in meiosis (Figure 8c). Typically, four pairs of chromosomes are present in the hybrid, together with a ring of four chromosomes and, for the *biloba* and *australis* subspecies, a chain of five chromosomes (including an extra chromosome from *C. lingulata*). The chain of five indicates that the additional chromosome of *C. lingulata* is homologous with parts of two chromosomes that are common to both these subspecies. In the hybrid between *C. lingulata* and *C. b. brandegeae*, instead of a chain of five, a chain of three and an additional pair is formed at metaphase. At anaphase (Figure 8d), the stage when chromosomes move to the opposite poles, a bridge (i) is present in the hybrids, indicating an inversion in one of the other chromosomes (Figure 6c). The appearance of the chromosomes during meiosis as shown in Figure 8 is interpreted in Figure 9.

26

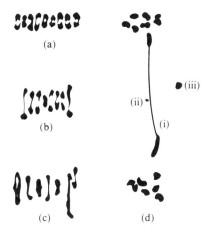

(a)

(b)

(c)

(d)

(i) (ii) (iii)

FIGURE 8 Appearance of chromosomes during meiosis. (a) *Clarkia lingulata*, metaphase I; (b) *C. biloba australis*, metaphase I; (c) hybrid, metaphase I, showing four pairs of chromosomes, a ring of four (left) and a chain of five (right); (d) hybrid, anaphase I, showing (i) a bridge, (ii) a fragment, and (iii) a lagging chromosome.

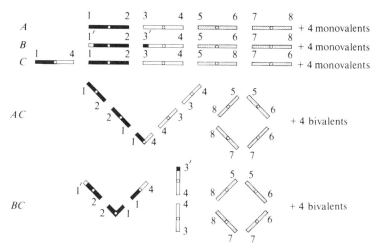

FIGURE 9 The structural and quantitative relationships of the haploid genomes of *Clarkia biloba* subspecies and *C. lingulata*. (A) Genome of *C. biloba australis* and *C. b. biloba*; (B) genome of *C. b. brandegeae*; (C) genome of *C. lingulata*. AC and BC are hybrids of A and C, and B and C, respectively, as seen at meiotic metaphase I. The numbers distinguish different chromosome arms. In hybrid AC, the chain of five chromosomes and ring of four chromosomes are formed by pairing between different chromosomes. The primes (') on some arms indicate that these arms have been slightly modified.

Figure 9 indicates that the genome of *C. lingulata* is structurally more similar, and hence more closely related, to that of the subspecies *australis* and *biloba* than it is to *brandegeae*. It is clear that the degree of reproductive isolation is not a reliable indicator of genetic relationship, because hybrids of *C. lingulata* with *brandegeae* are more fertile on average than those with either *biloba* or *australis* (even though the hybrids themselves are more difficult to obtain). The reason is that the frequent formation of a chain of five chromosomes in hybrids of *C. lingulata* with *biloba* or *australis* (AC in Figure 9) causes more disturbance to the assortment of chromosomes, and results in a much greater incidence of inviable gametes, than hybrids with *brandegeae* (BC in Figure 9).

H. Lewis has presented a hypothesis for the origin of new species in *Clarkia* which involves the following stages:

1 An exceptional drought reduces a normally outcrossing population to a few plants.

2 The few survivors or founders re-establish the population by undergoing self-pollination.

3 Extensive chromosome rearrangements occur and cause structural heterozygotes to be partially sterile.

4 A chromosomally monomorphic population is formed by chance from a homozygous combination of rearranged chromosomes.

5 Isolation of the new species from the parent is ensured both by hybrid sterility and by self-fertility.

The process is envisaged to be rapid and has been referred to as *saltational speciation*. In the case of *C. lingulata* and *C. biloba*, extensive field, garden and laboratory experiments suggest that the former is no better adapted to the marginal environment it occupies than the surrounding populations of the latter. Hence, Lewis believes that the origin of *C. lingulata* was largely accidental, and the new species has no adaptive superiority to the parental one. The inferred relationship between *C. lingulata* and *C. biloba* is supported by electrophoretic studies which have shown the close similarity between certain enzymes of the two species.

4.1.3 Karyotype evolution in mammals

The karyotypes of about 1 500 species of mammals have now been studied. A great diversity of karyotype exists in most orders, and there seems to have been a far greater tendency for mammalian speciation to be accompanied by major chromosomal rearrangements than speciation in other groups of animals. A. C. Wilson and his co-workers have estimated the number of karyotype changes per lineage per million years (i.e. the rate of chromosomal evolution) in various groups of placental mammals, marsupials and other vertebrates and also calculated a total speciation rate for each of the same groups.

When they plotted total speciation rate against rate of chromosomal evolution for the major vertebrate groups they obtained a highly significant correlation. This very strongly supports the view that total speciation rate and the rate of chromosomal evolution in the major groups of vertebrates are related. Their calculations also show that the rate of evolution, measured by total speciation rate, is higher in mammals as a whole than for other vertebrate groups. And among the mammals, horses and primates top the list, while bats and whales evolved as slowly as most other vertebrates.

Based on these results, G. L. Bush and his co-workers suggested that:

1   speciation and chromosomal evolution in mammals are causally related;

2   speciation and chromosomal evolution seem to take place most quickly in those groups where there is social structuring (such as clans or harems) or other factors which reduce adult *vagility* (the range of mobility connected with reproduction).

vagility

□ What are the possible consequences of social structuring, territoriality and other factors which reduce adult vagility in mammals?

■ These factors could have important effects on the breeding structure of the population, encouraging the subdivision of the population into small temporally stable local groups, where the founder effect, associated with the fixation of particular alleles or of a particular chromosome rearrangement, can bring about rapid speciation without strict geographic isolation. The reduced fertility of karyotypic heterozygotes acts as an effective barrier between local groups, accelerating genetic divergence.

Bush's hypothesis predicts that effective population size should be *inversely* related to both the rate of speciation and the rate of chromosomal evolution. Unfortunately there have been few estimates of effective population size in the different taxa.

4.2   Genetic differences between species

Genetic differences between species have been investigated using three main techniques: electrophoresis, nucleic acid hybridization and immunological procedures. We shall discuss each of them in turn.

4.2.1   Electrophoresis

In Section 3.2.2 we considered genetic divergence in *Drosophila* at different levels of evolutionary differentiation calculated from electrophoretic data. Since the work of Ayala and his colleagues, numerous studies have been carried out on other groups of species. A summary of these findings is given in Table 7.

TABLE 7 Average genetic distance, $\bar{D}$, for several taxa at various levels of speciation

Species	No. of loci	No. of taxa	$\bar{D}$
Local populations			
Drosophila willistoni group	36	—	0.031
Drosophila equinoxialis	27	—	0.026
Drosophila mulleri	16	—	0.002
Homarus americanus	44	—	0.006
Peromyscus boylii and pectoralis	23	—	0.030
Subspecies			
Drosophila willistoni group	36	4	0.230
Drosophila equinoxialis	27	2	0.255
Drosophila mojavensis	16	2	0.130
Lepomis macrochirus	18	2	0.171
Taricha torosa	18	3	0.029
Peromyscus boylii and pectoralis	23	10	0.052
Sibling species			
Drosophila mulleri-aldrichi	16	2	0.123
Drosophila willistoni group	36	6	0.581
Thomomys talpoides complex	31	6	0.078
Peromyscus truei group	23	4	0.057
Semispecies			
Drosophila willistoni group	36	6	0.226
Spermophilus mexicanus-tridecemlineatus	28	2	0.036
Neotoma sp.	20	3	0.123
Peromyscus leucopus-gossypinus	23	2	0.178
Well differentiated species			
Drosophila willistoni group	36	6	1.056
Lepomis sp.	14	10	0.609
Taricha sp.	18	3	0.466
family Icteridae	15	7	0.227
Geomys sp.	17	5	0.470
Peromyscus boylii group	17	5	0.323

☐ From the data in Table 7, is the amount of genetic difference, by itself, a useful criterion for determining the evolutionary status of populations of organisms?

■ It is quite clear that genetic difference, at least as measured by electrophoresis, cannot be used as a criterion for determining the evolutionary status of different organisms. Well differentiated species in different families show a wide range of genetic distances (from 0.227 in the family Icteridae to 1.056 in the Drosophila willistoni group), and these overlap with the subspecies genetic distances (0.029 to 0.255). Subspecies, semispecies and even sibling species are virtually indistinguishable in terms of their ranges of genetic distance. Within single genera, the amount of genetic difference may show a gradation correlated with taxonomic status (e.g. in the genus Peromyscus). However, even here, the criterion often breaks down as within the genus Drosophila (see Section 3.2.2.)

4.2.2 DNA hybridization

When two organisms differ genetically, this difference is expected to reside in the base sequence of their DNA. DNA exists as two complementary strands, so that if these are dissociated in vitro they reassociate more or less completely. When DNAs from different species are dissociated together in a mixture and then allowed to reassociate, hybrid DNA can be obtained in varying degrees depending on the amount of similarity (or homology) between the base sequences in the two DNA species. Some strands that associate will not be completely complementary in base sequence and these strands will tend to dissociate again when the temperature is raised. So the degree of mismatch between the nucleotides can be estimated from the temperatures at which hybrid strands come apart.

In order to measure homology in DNA base sequence between two species, A and B, DNA from one species, A (Figure 10a), is denatured or 'melted' into single stranded DNA by heating it to 100 °C (Figure 10b) after which it is cooled rapidly so that the strands remain separate. The single stranded DNA is then trapped in a homogeneous matrix of Agarose or nitrocellulose membrane filters, which is then chopped into small pieces. This breaks up the DNA into single stranded fragments. A known amount of this sheared, single stranded DNA is incubated with a known amount of radioactively labelled single stranded DNA from the same species A, plus a known amount of single stranded DNA from the second species B (Figure 10c). The mixture is incubated at 60 °C for several hours to allow association between the free DNA and the DNA trapped in the filter fragments (Figure 10d).

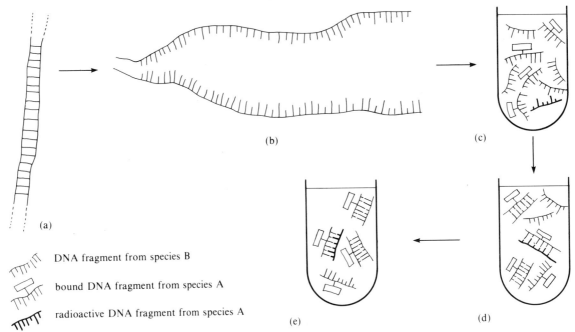

DNA fragment from species B

bound DNA fragment from species A

radioactive DNA fragment from species A

FIGURE 10 Measuring DNA base sequence homology.

The remaining (unassociated) DNA is then washed from the filter fragments (Figure 10e) and the amounts of DNA bound from species A and B can then be determined. The procedure is repeated, with larger and larger amounts of single stranded DNA from species B.

If sequence homology exists between species A and species B, there will be competition for association with the trapped DNA fragments between free DNA from species B and the free radioactively labelled DNA from species A. As the amount of DNA from species B is increased, the association of radioactively labelled DNA will decrease to a constant value when all the homologous specific associations have taken place. This gives a direct estimate of the percentage of sequence homologies in the two species.

The result of a series of typical experiments is shown in Figure 11. Denatured DNA from *Drosophila melanogaster* was trapped by the filter and mixed with $1 \mu g$ (10^{-6} g) of tritiated (^{3}H) DNA from this species. Then increasing amounts of non-radioactive DNA from two other species, *D. funebris* and *D. simulans*, separately competed for association with the trapped DNA. A control was set up in which the non-radioactive competing DNA was from *D. melanogaster* itself. As the amount of unlabelled DNA from *D. melanogaster* increased, the amount of bound radioactive DNA decreased and gradually approached zero. In contrast, the maximum amount of DNA bound from *D. funebris* was only 20 per cent, and that from *D. simulans* about 80 per cent, despite a continued increase in the amount of DNA added. This shows that about 80 per cent of DNA from *D. funebris* and 20 per cent of DNA from *D. simulans* had no sequence homology with *D. melanogaster*. It follows that the homology of *D. funebris* to *D. melanogaster* is 20 per cent whereas that of *D. simulans* is 80 per cent. Results obtained with this kind of experiment, using human DNA, are presented in Table 8.

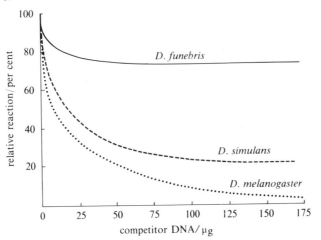

FIGURE 11 Homology between the DNA sequences of *Drosophila melanogaster* and those of two other species, as shown by the relative reaction of labelled *melanogaster* DNA in the presence of competitor DNA. (From *Evolution* by Theodosius Dobzhansky, F. J. Ayala, G. Leyadrs Stebbins and James W. Valentine. W. H. Freeman and Company. Copyright © 1977.)

TABLE 8 Competition between 200 μg (or more) of DNA of various species and 0.5 μg of ^{14}C-labelled human DNA for binding with 0.5 μg of human DNA held on agar

DNA tested	Degree of taxonomic differentiation	Inhibition of human–human DNA binding/per cent
human	—	100
chimpanzee	family	100
gibbon	family	94
rhesus monkey	superfamily	88
capuchin monkey	superfamily	88
tarsier	suborder	65
slow loris	suborder	58
galago	suborder	58
lemur	suborder	47
tree shrew	suborder	28
mouse	order	21
hedgehog	order	19
chicken	class	10

☐ What is the estimated sequence homology (a) between human and chimpanzee DNA, and (b) between human and mouse DNA?

■ (a) 100 per cent; (b) 21 per cent.

4.2.3 Immunological distances

Estimates of the degree of similarity between proteins from different species can be obtained by immunological techniques. A protein such as albumin is purified from a species, say a human, and is injected into another species, typically a rabbit. The rabbit produces antibody proteins (called immunoglobulins) against the foreign protein (called the antigen). These antibodies react with the antigen, combine with it and precipitate it out of solution. Antibodies are very specific to the antigen used, but proteins homologous to the original antigen, such as albumin from related species, will also react with the antibody to a greater or lesser degree depending on its similarity to the original antigen. The degree of dissimilarity between a protein from one species and a similar protein from another species is expressed as immunological distance, and is approximately proportional to the number of amino acid differences between the homologous proteins.

Immunological distances between humans, apes and Old World monkeys are given in Table 9. Separate antibodies were prepared against albumin obtained from humans, chimpanzees and gibbons, isolated and reacted with albumins from six species of apes and six species of Old World monkeys. The tests with antibodies prepared against humans show that albumins from the African apes (chimpanzee and gorilla) more closely resemble those from humans than the albumins of Asiatic apes (orang-utan, siamang and gibbon). The albumins from the Old World monkeys differ most from the human ones. The antibodies to chimpanzee albumin gave similar results to the antibodies to human albumin. The Table indicates that the albumins of gibbon and siamang are very similar; it also shows that there is not much more difference between orang-utan albumin and the albumins of other Asiatic apes than there is between the albumins of the African apes.

TABLE 9 Immunological distances between albumins of various primates

Species tested	Antibodies to albumin from		
	human	chimpanzee	gibbon
human	0	3.7	11.1
chimpanzee	5.7	0	14.6
gorilla	3.7	6.8	11.7
orang-utan	8.6	9.3	11.1
siamang	11.4	9.7	2.9
gibbon	10.7	9.7	0
Old World monkeys (average of six species)	38.6	34.6	36.0

31

Another protein, lysozyme, gave results not entirely consistent with those for albumin. On the lysozyme data, the gorilla is further from humans than the orang-utan is. Inconsistencies like this indicate that inferences about phylogenies based on a single protein may be quite misleading, because different proteins may have evolved at different rates in different lineages. Thus mosaic evolution (Unit 8) may occur at the molecular level as well as at the morphological level.

A refinement of the method for comparing single homologous proteins is by amino acid sequencing. Again, consideration of a single protein rarely gives a clear indication of the true phylogenies of the species involved.

4.3 The relationship between organismal and genetic differences between species

Studies on interspecific differences such as those described in the last Section lead to questions concerning the relationship between organismal and genetic differences between species. Organismal differences are those involving the biology of *whole* organisms, whereas genetic differences are those involving differences in specific genes which code for proteins. We have already alluded to the discordancy between organismal and genetic change in Section 4.3 of Unit 8. Here we shall describe more instances of this.

organismal differences
genetic differences

A. C. Wilson and his colleagues compared the rate of evolution in two vertebrate lineages—frogs and placental mammals. Placental mammals have experienced much more rapid organismal evolution than the lower vertebrates. Although there are thousands of frog species living today, they are so uniform phenotypically that zoologists place them into a single order (Anura), whereas placental mammals are divided into at least 16 orders. The anatomical diversity represented by bats, whales, cats and horses is not paralleled among frogs (see Table 10).

TABLE 10   Rates of evolution in frogs and placental mammals

Property	Frogs	Placental mammals
number of living species	3 050	4 600
number of orders	1	16–20
age of the group/Ma	150	75
rate of organismal evolution	slow	fast
rate of albumin evolution	standard	standard
rate of change in chromosome number	slow	fast
rate of change in number of chromosomal arms	slow	fast

A comparison of immunological distances (estimated from the albumins of hundreds of species of frogs and mammals) has shown that frog species can differ by as much as a bat differs from a whale. The β-polypeptides of haemoglobin from two species of *Rana* differ by 29 amino acid substitutions, a greater difference than found between any two orders of placental mammals. The difference in nucleotide sequence of DNA between the subspecies of Anurans such as *Xenopus laevis laevis* and *Xenopus laevis borealis*, is greater than that between humans and New World monkeys.

Frogs with widely different proteins show little difficulty in hybridizing, whereas even closely similar species of mammals usually fail to give viable hybrids. Thus if one assumes that the degree of species isolation as measured by the production of inviable hybrids is an indication of organismal evolution, organismal evolution has gone much further in mammals than in frogs.

Another discordance between the apparently different rates of evolution in frogs and mammals is seen in chromosomal changes. Placental mammals have experienced far more rapid karyotypic changes than frogs (see Table 10) in contrast to the *slowness* of protein changes in the mammals. This is seen in Figure 12, where the percentage of pairs of species having the same chromosome number is plotted against the immunological distance between the albumin of the pairs, for both frogs and mammals.

ITQ 9   What is the approximate immunological distance (measured for albumin) at which there is a 50 per cent chance that two species will differ in chromosome number in (a) frogs, and (b) placental mammals?

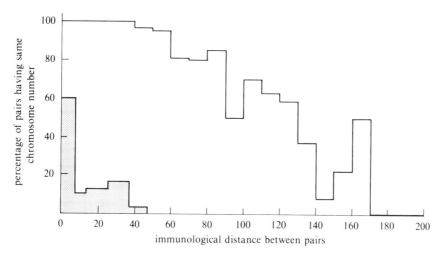

FIGURE 12 Proportion of species pairs of identical chromosome numbers, as a function of the immunological distance between the albumins of the pairs. The white histogram summarizes the results for 373 different pairs of frog species, and the grey one summarizes the results for 318 different pairs of placental mammal species.

These results highlight a major problem in the study of evolution: the exact relationship between organismal and genetic change. (The term genetic is ambiguous in this context, a more precise one is *genic*, which emphasizes that specific genes coding for proteins are involved. But as the electrophoretic surveys were done on the assumption that they *are* representative of genetic change, we shall use 'genetic' here.) With respect to their proteins when examined electrophoretically, human and chimpanzee are as close as sibling species genetically, yet they are classified in different families on anatomical and physiological grounds. The contrast is brought out in Figure 13. The rates of genetic change in both lineages are the same, yet in humans organismal change has far outstripped genetic change.

If genetic change is not related to organismal change, then the current neo-Darwinian theory of evolution, which attempts to explain evolution solely in terms of the substitution of alleles in populations, must be incomplete at best. One way out of this difficulty is, as A. C. Wilson points out, to ascribe the accelerated change in the human lineage to a few *regulatory genes* which have no identified gene products. So far all studies on genetic distances have been done on *structural genes* (i.e. genes that code for protein), so the possibility remains that drastic changes in other genes have gone undetected. But Wilson goes further and suggests that the rapid chromosomal evolution in placental mammals is the basis of changes in genetic regulatory systems. It seems unlikely, however, that chromosomal rearrangement as such has anything to do with changes in genetic regulatory systems, because many species and species groups show a considerable amount of karyotypic variation without a corresponding variation at the organismal level. Conversely, widely divergent species may have very similar karyotypes. For example, the four living species of Hominoidea (*Pan troglodytes, Gorilla gorilla, Pongo pygmaeus* and *Homo sapiens*) are quite different morphologically and physiologically, but they have very similar karyotypes. There has been a minimum of chromosomal rearrangements in the line leading to *Homo sapiens*.

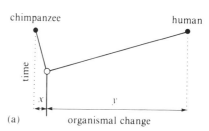

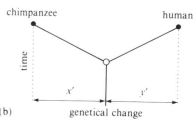

FIGURE 13 The contrast between organismal evolution and genetical evolution since the divergence of the human and chimpanzee lineages. (a) More organismal change has taken place in the human lineage (y) than in the chimpanzee lineage (x). (b) Protein and nucleic acid evidence indicates that as much change has occurred in chimpanzee genes (x') as in human genes (y').

4.4 Are species differences adaptive?

We have seen that species differ in morphology, in physiology, in chromosomal constitution and in genetic make-up. We must somehow relate these differences between species to the different ecological niches of the species; in other words, we need to show that at least some of the morphological, physiological and genetic attributes characterizing a species are adaptive to some aspect of the species' niche.

We have seen that, *within* a species, the geographical variation of physiological characters can in general be related to broad differences in physical and climatic conditions. Hence we are likely to find that, when a number of species co-exist in the same habitat, *parallel* adaptations exist in some of the species to factors such as the temperature, humidity and soil type found there. (Indeed, there may well be adaptations to factors in the microhabitat which at present we know nothing about.)

The adaptedness of a newly evolved species, *Clarkia lingulata*, was compared with its parent species, *C. biloba*, and no adaptive superiority of the new species in its native environment could be demonstrated. This emphasizes the point made by many evolutionists—including Darwin himself—that not all evolutionary changes can be understood in terms of adaptation. Certainly, speciation does not necessarily involve adaptation; and conversely, obvious adaptations such as trophic specializations can occur without speciation. Adaptive processes, however, can be enormously speeded up by factors favouring the formation of new species. The most striking examples are the adaptive radiations which often accompany the evolution of whole flocks of species (see Section 5.2). Thus, for the moment, we may conclude that:

1   Differences between species may or may not be adaptive. This is not to deny that each species is adapted to its environment, but only implies that the different species can represent alternative solutions to the same or similar problems in survival. (In the television programme 'The picture wings of Hawaii' you will see examples of species which are reproductively isolated but live in identical habitats and apparently have no adaptive differences between them.)

2   Species may be viewed as being 'co-adapted' gene pools in which genes have been co-selected for their harmonious functioning. Thus genetic differences between species are unlikely to be accounted for in terms of a given number of allelic substitutions.

3   The interdependence of gene action means that genes will have pleiotropic or multiple effects. Selection of one aspect of a gene function will often carry along other effects of no adaptive value.

4   Not all change is adaptive. The frequent involvement in speciation of genetic drift or of founder effects in small populatons (see Section 5.1.2) is perhaps the strongest circumstantial evidence that speciation itself is not primarily adaptive even though the speciation process may give rise to circumstances favourable to the evolution of special adaptations.

4.5   Objectives and SAQs for Section 4

Now that you have completed this Section you should be able to:

(a)   Reconstruct possible phylogenies from data about chromosomal mutations.

(b)   Describe the relationship between the rate of chromosomal evolution and the rate of speciation, in mammals and other vertebrates.

(c)   State the relationship between genetic change and organismal evolution.

To test your understanding of this Section, try the following SAQs.

SAQ 5 (*Objective a*)   Three species of *Drosophila* on Sun Island have the following banding sequence in one of their chromosomes:

Species 1:   A B C D E F G H

Species 2:   A E D C B F G H

Species 3:   C A B D E F G H

A single species inhabits the nearest island, Moon Island, and has the following banding sequence in the same chromosome:

Species 4:   A G F B C D E H

Geologically, Moon Island is 500 000 years older than Sun Island. Work out the phylogeny of the four species.

SAQ 6 (*Objectives b and c*)   (a)   Summarize the evidence in favour of a causal relationship between chromosomal mutations and speciation.

(b)   Recapitulate the arguments leading up to the hypothesis that regulatory gene mutations played a part in the evolution of *Homo sapiens*.

5 Speciation

Study comment This Section considers a model of speciation based on geographical isolation, and the evidence supporting it. The implications of the model for neo-Darwinian theory and the interpretation of the fossil record are briefly discussed.

The orthodox view of species formation, derived from Darwin's belief in the gradual nature of evolutionary change, envisages a continuum of increasing genetic divergence as one moves from populations to races, to subspecies and finally to reproductively isolated species. Recent evidence, however, indicates a general lack of correlation between genetic divergence as indicated by molecular studies and the organismal changes associated with the formation of new species (see Section 4.3). The solution to this paradox lies in the precise role played by geographical isolation in speciation.

Traditionally, genetic differences between populations were regarded as being mainly the accumulated results of the action of natural selection. Geographical isolation was regarded merely as a secondary factor which *facilitated* the accumulation of genetic difference; once these differences reached a critical amount, the population was transformed into two reproductively isolated species.

Recently, geographical isolation has been increasingly seen by many as the most important factor in speciation. Once geographical isolation is taken into account, as R. Goldschmidt pointed out, the process no longer has to be gradual. This is because any isolated population can rapidly evolve into a different species, regardless of whether it was originally a subspecies or a local population. In other words, the *initial* amount of genetic divergence is quite irrelevant. Recent evidence is not incompatible with this view.

By definition, a new species is formed once reproductive isolation is established. But reproductive isolation, as we have seen in previous Sections, may occur with or without appreciable genetic change. To account for speciation, we therefore have to account for the origin of reproductive isolation between populations.

5.1 Allopatric speciation

Allopatric speciation refers to the formation of new species in populations geographically isolated from the parent species. The general model put forward by Ernst Mayr in 1957 involves the following stages:

1 Geographical isolation. A species is split up by some geographical barrier into more or less isolated populations. This split can occur either in the main range of distribution of the species or at the margins of the range (*peripheral isolates*).

2 Genetic reorganization and divergence. This takes place in the isolated population, often but not necessarily accompanied by a change in habit or habitat of the organism.

3 Reproductive isolation. This develops chiefly as a by-product of genetic divergence.

Species may go through several cycles of this sequence of events before reproductive isolation becomes complete. In Sections 5.1.1 to 5.1.3 we consider Mayr's three stages in turn.

5.1.1 Geographical isolation

According to Mayr (and many of his predecessors), geographical isolation is an essential prerequisite to speciation. This is because there must be some means of preventing genetic exchange (gene flow) between the population and the parent species, so that a genetic combination (genome) adapted to local conditions can be built up.

There has been some disagreement over the role of geographic isolation in speciation. Some workers contend that it is not always a prerequisite to speciation. New species could arise *sympatrically*, that is, in the same geographic area without isolation. We shall examine the role of the various factors which can result in sympatric speciation in the next Section.

sympatric speciation

35

Geographic isolation depends on the creation of islands of populations between which gene flow is prevented. Such islands can arise through the establishment of barriers, such as the following:

(a)   The sea. The effectiveness of the sea as an isolating agent for terrestrial organisms was fully appreciated by Darwin, Wallace and others. The finches of the Galapagos Islands and the drosophilids in the Hawaiian Islands (see the television programme 'The picture wings of Hawaii') are classic examples of rapid and profuse speciation through isolation on islands.

(b)   Mountain ranges. These are important barriers, particularly if they separate climatic zones. For example, the Himalayas separate India from Tibet.

(c)   Valleys. Just as mountains are barriers for lowland organisms, valleys are barriers for mountain species.

(d)   Glacial masses. The Pleistocene ice masses of the northern continents were among the most effective barriers in recent times. In Europe, the Scandinavian icecap and the Alpine glaciers came within 330 km of each other, separated merely by a cold steppe that formed an effective barrier between the unglaciated Atlantic region and areas along the eastern Mediterranean and in the Near East. This led to much subspeciation and possibly some speciation.

(e)   Zonation of vegetation. The borders of vegetation belts are often very sharp in the tropical and subtropical regions, and form effective geographic barriers for many animals. This has been shown for birds in Africa, Australia and Amazonia.

(f)   Distribution of freshwater. Bodies of freshwater, such as streams, rivers and lakes, are well isolated. On the whole, each stream or stream basin is a population unit separated by land from adjacent streams. Lakes are for water organisms what islands are for land organisms. Every ancient freshwater lake (e.g. Lake Baikal and Lake Tanganyika) has its own endemic fauna. Furthermore, each lake consists of an archipelago of areas (such as a rocky shores), each a habitable 'island', separated from other similar areas by barriers of unsuitable habitats (such as sandy or muddy shores).

Apart from the above 'extrinsic' factors of geographic isolation, Mayr lists two 'intrinsic' factors influencing the effectiveness of the geographic barriers. These concern the physiological and behavioural properties of the organisms, and are as follows:

extrinsic and intrinsic factors of geographic isolation

(a)   The effectiveness of dispersal. Many aquatic organisms have pelagic larval stages which facilitate dispersal over wide areas. Such organisms are cosmopolitan in distribution and show little or no geographical differentiation into races. Examples are rotifers and freshwater crustaceans. Animals differ greatly in their propensity for active dispersal, and the factors influencing this propensity are not well understood. Plants too show marked differences in the effectiveness with which they disperse their seeds or vegetative propagules.

(b)   Behavioural reinforcement of geographic isolation. Even in animals with high mobility, it often happens that there are aspects of mating behaviour which reinforce geographic isolation. These include homing and territoriality in vertebrates, especially social structuring in mammals (see Section 4.1.3). Habitat preference strongly reinforces geographical isolation, and in some cases it may be the prime cause of speciation (see Section 6.1). In the television programme 'The picture wings of Hawaii', you can see the crucial part behaviour plays in the evolution of Hawaiian drosophilids. Where species show geographical or ecological isolation, sexual behaviour acts to reinforce this.

It must be emphasized that geographic barriers are not static entities, they have an evolution of their own. Sea-levels rise and fall—and islands form, join with other islands or disappear. Glaciers advance and retreat, streams and lakes likewise swell or become dry. Climatic fluctuations cause the contraction and expansion of vegetation zones. In short, as we emphasized in Unit 9, a whole dynamic interplay of physical and biotic factors, now effecting isolation, now causing emigration or secondary contact and hybridization, contribute to the complexity of the speciation process and the seemingly infinite variety of living forms.

5.1.2   Genetic reorganization and divergence in isolated populations

This is one of the more controversial areas in Mayr's model. His argument runs as follows:

1   The isolate represents a restricted sample of the genotypes of the species (founder effect) so that a genetic background (which is different from that of the species as a whole) is established against which individual genes alter in their selective value.

2   The restricted sample leads effectively to increased inbreeding and homozygosity (see Unit 10) and alleles will be selected that function best in a homozygous state.

3   The restriction of gene flow between the isolated population and the rest of the species enables new combinations (genotypes) to be built up that are adaptive to the local environment. This contributes to an *ecological shift*—the adaptation to a new ecological niche distinct from that of the parent species.

ecological shifts

4   Natural selection is relaxed following (i) a reduction in competition (density dependent selection, see Unit 11), and (ii) the operation of density independent selection due to physical and biotic factors. In marginal environments this tends to speed up the ecological shift and genetic changes by causing wide fluctuations in population size.

5   In small populations, *founder effects* and *random genetic drift* cause the rapid *fixation of genes*, increasing the genetic divergence of the population from the parent species.

fixation of genes

The combined effect of these five processes is a major reorganization of the genome which Mayr called a *genetic revolution*.

genetic revolution

Based on the above, at least two predictions can be made:

(a)   Isolated populations will evolve more rapidly on average than populations in the main range of a species.

(b)   Isolated populations will be genetically less heterogeneous than those in the main range.

Some evidence supporting both (a) and (b) exists, for instance: (i) The morphological and physiological characters of peripheral isolates are frequently more distinct than those of contiguous populations in the main range of the species, as for example in certain kingfishers (the *Tanysiptera hydrocharis-galatea* group in New Guinea). (ii) Electrophoretic studies of protein polymorphisms in 24 loci of *Drosophila pseudoobscura* show that, though there is much polymorphism, there is also a remarkable uniformity in the frequency of various alleles over the main distribution area of the species, with little or no differentiation into races. One isolated population, however, was genetically much less heterogeneous both in terms of the number of polymorphic loci and in average heterozygosity. Similar observations of reduced genetic diversity in island populations as compared with mainland populations have been made for numerous other groups of organisms.

What remains in doubt is whether genetic revolution is an essential part of speciation. We do not know how to identify a revolution of this sort, since the present studies of genetic differentiation rely almost entirely on the products of structural genes, telling us very little about the organization or reorganization of the genome.

5.1.3   Origin of reproductive isolation

Mayr favours the view that reproductive isolation is purely a by-product of genetic divergence. Others, including T. Dobzhansky, favour the view that isolating mechanisms are built up through natural selection when two incipient species begin to become sympatric, that is when their geographic ranges begin to overlap. The latter view is based on the observation that hybrids between two species are usually of lowered fitness. The argument is that individuals with inefficient isolating mechanisms (especially of the pre-mating type) will be susceptible to hybridization in areas of contact between the parental species and incipient new species. These genotypes will be eliminated from both populations as a consequence of selection against the hybrids they produce. This selection arises because the hybrids are less well adapted than either parent to their respective habitats. Genotypes with better-developed pre-mating isolating mechanisms are not likely to hybridize and will be preserved by natural selection. The

Mayr's view
Dobzhansky's view

major flaw in this theory is that it *assumes* that hybrids are selected against in the first place, that is, it proposes to explain how reproductive isolation arises by assuming it already exists! It could account for the *reinforcement* of reproductive isolation by selection for pre-mating mechanisms once post-mating mechanisms have been established. The mechanism certainly cannot explain the cases where pre-mating isolation exists in the absence of any reduction in hybrid fitness.

☐ Which of the following would be evidence (a) in support of, (b) against, and (c) neither for nor against the theory that reproductive isolation is the result of natural selection?

(i) No difference exists between the sympatric and allopatric populations of species with respect to the degree of reproductive isolation.

(ii) The sympatric populations of different species are more strongly reproductively isolated than allopatric populations.

(iii) The sympatric populations of different species are less strongly reproductively isolated than allopatric populations.

■ (a) (ii); (b) (iii); (c) (i).

As a matter of fact, *all* of (i)–(iii) are found in species of *Drosophila*, with (i) being the most frequent.

Another piece of evidence often cited in support of this theory is the phenomenon of *character displacement* (see Unit 11). This refers to the more pronounced differences in morphological characters in the sympatric populations of two species compared with those found in the allopatric populations. Here again, instances where the reverse is true also exist in nature, and no definite conclusions can be drawn. Moreover, selection for character displacement may simply be selection for specialization and is not necessarily connected with selection against hybrids.

A third line of evidence in favour of this theory are the cases where hybridization occurred freely in historical times, but it now occurs much less freely or not at all. An example is the invasion of the woodpecker *Picoides syriacus* into the range of another woodpecker, *P. major*, in southwestern Europe. At first, quite a number of hybrids were reported, but only a few have been noted subsequently except along the actual line of expansion of *P. syriacus*.

Cases where the frequency of hybridization has not changed appreciably in recent times are also well known. For example, the mallard (*Anas platyrhynchos*) and the black duck (*A. rubripes*) are sympatric in the eastern United States and no change in the frequency of hybridization has been observed for the past 75 years. The most convincing argument against natural selection being the main or sole factor in bringing about reproductive isolation is supplied by the old hybrid belts which have existed in many cases for thousands of years. The narrowness of these belts indicates that the hybrids are selected against, but there are no indications that isolation has been strengthened in any of the cases studied.

Mayr's theory is consistent with all these observations, simply because it does not make definite predictions about when and where reproductive isolation develops. All it states is that reproductive isolation is a chance by-product of genetic reorganization and divergence between populations. As we have indicated above, there is as yet no way to measure genetic reorganization although the amount of genetic divergence in structural genes is more or less known for many cases. *Certainly no consistent correlation has so far been shown between reproductive isolation and genetic divergence.*

5.2 Species flocks

Species flocks or *species swarms* are large groups of closely related species confined to a narrow geographic area, often with no *close* relatives elsewhere. Detailed studies of species flocks, involving the geological history (both physical and biotic) of the environment, and the biology, life histories and genetic relationships of the organisms, make possible the reconstruction of the sequence and mode of speciation. We shall study two examples in this Course: one, the cichlid fishes of Lake Victoria in East Africa, will be described in the next Section, and the other, *Drosophila* species of Hawaii, is the subject matter of the television programme 'The picture wings of Hawaii'.

species flocks or swarms

5.2.1 The cichlid fishes of Lake Victoria

The cichlids dominate the fish fauna of Lake Victoria both ecologically and in terms of species numbers. There are over a hundred endemic species of the genus *Haplochromis* which have evolved within the past 0.75 Ma.

Haplochromis

Lake Victoria is a drainage basin fed by several large rivers (Figure 14). Its only outflow, the Victoria Nile, formed late in the history of the lake and established contact with a much smaller basin (Lake Kioga) in the north and so with the River Nile. The fauna of Lake Victoria has always been isolated from that of the Victoria Nile and Lake Kioga, originally by the Ripon Falls and more recently by a dam that replaced and submerged the Falls. Lake Victoria originated during the mid-Pleistocene, about 750 000 years ago, at which time the future lake basin was crossed by several westward-flowing rivers which drained the eastern highlands of Kenya. Later, as a result of geological changes to the west of what is now Lake Victoria, a two-way drainage system was established, eastwards into the developing Victoria basin and westwards into several protolakes to the west of Lake Victoria (these protolakes have now become Lakes Albert, George, Edward and Kivu).

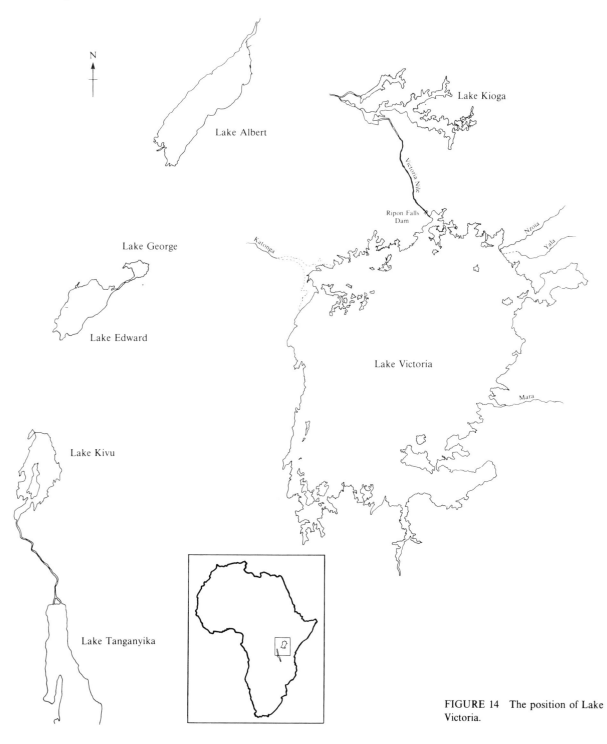

FIGURE 14 The position of Lake Victoria.

As the river valleys gradually filled, the area became a series of shallow lakes. Eventually, each of these joined its neighbour, linking together to form a single body of water that occupied an area greater than Lake Victoria does today.

Even after the single body of water came into existence, the lake basin was subject to periods of tectonic instability which caused changes in water level, leading to the isolation and later reincorporation of small peripheral bodies of water.

The Lake Victoria *Haplochromis* species flock may have originated from a single ancestral species or a group of related species which resembled the living species, *H. bloyeti*. The circumstantial evidence for this includes:

1 Lakes Victoria, Albert, Edward, George and Kivu were originally fed by the East African highland drainage system. The species contained in these lakes are similar in many respects, and they do not show certain features found in species forming the flocks of other East African lakes (e.g. Lake Tanganyika to the southwest of Lake Victoria) which were never fed by that drainage system.

2 All the relics of this drainage system are now populated by *H. bloyeti* (or by a number of species all virtually indistinguishable from *H. bloyeti*). However, *H. bloyeti* does not occur in neighbouring lakes that never formed part of that drainage system.

3 The other, more generalized, species within the species flock resemble *H. bloyeti* and its close relatives.

The predominant adaptive radiation in the *Haplochromis* species flock is trophic specialization. Every major food source in the lake, except zooplankton, has been exploited by one or more *Haplochromis* species, see Figure 16. The major trophic groups include insectivores, crustacean eaters, plant eaters, mollusc eaters and paedophages (species that prey on the larvae of other fishes, mainly of other *Haplochromis* species). One species, *H. welcommei*, has the peculiar habit of feeding on fish scales, which it scrapes off the tails of other fishes. All groups, with the exception of most of the insectivores, exhibit specialized features connected with their feeding habits. Within each trophic group, there are several species which show a gradient of increasing specialization in their morphological features, especially those associated with their feeding habits. These morphological changes are mainly confined to the shape and size of dentition on the jaws and pharyngeal bones, dental patterns, and correlated changes in skull form (see Figure 15).

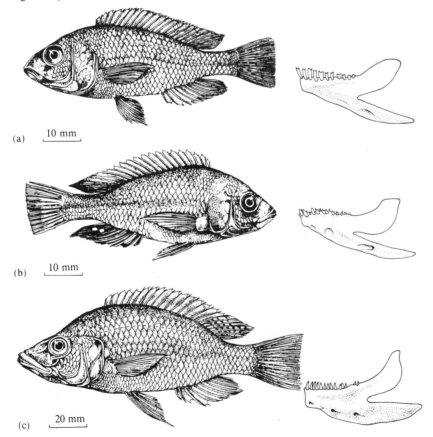

(a) 10 mm

(b) 10 mm

(c) 20 mm

FIGURE 15 Three *Haplochromis* species, showing the whole fish (left) and the lower jaw (right). (a) *H. pallidus*; diet–detritus an insects. (b) *H. ishmaeli*; diet–molluscs. (c) *H. parvidens*; diet–fish eggs and larvae.

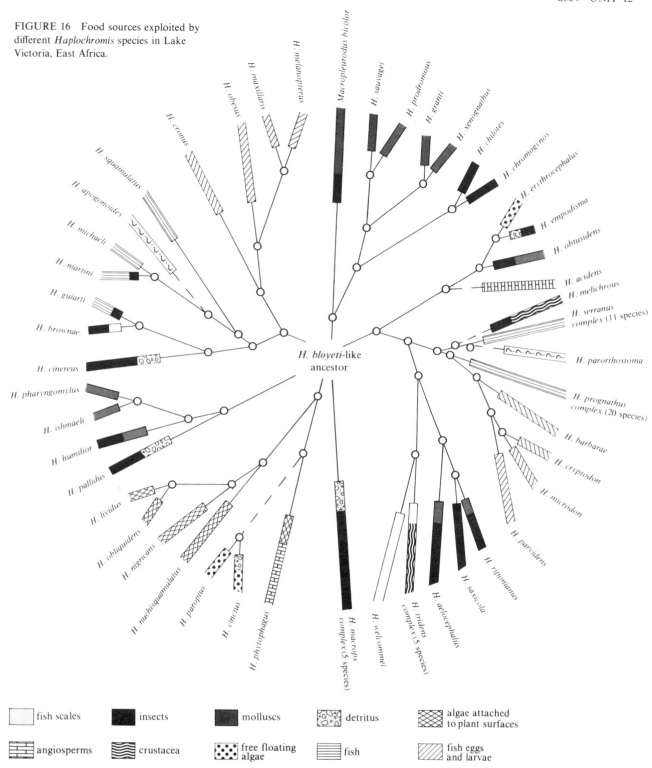

FIGURE 16 Food sources exploited by different *Haplochromis* species in Lake Victoria, East Africa.

In general, there is considerable interspecific overlap in feeding habits and habitats among the *Haplochromis* species of Lake Victoria. There is little apparent geographic restriction on the distribution of any species within the lake, and when geographic differentiation of populations within a species occurs, the morphological differences between the populations are very slight.

The females of *Haplochromis* species in Lake Victoria are probably all mouth brooders, i.e. the ova are fertilized and incubated in the mouth of the female. All known *Haplochromis* species are sexually dimorphic, the male always having the brighter and more colourful livery even in non-breeding periods. Interspecific differences in colours are not always sharp. A very prominently coloured feature in the males of all *Haplochromis* species (and related genera) are the ocelli (spots that resemble eggs, see Figure 15b) on the anal fin. After the female has laid eggs in a nest on the bottom she takes the eggs into her mouth. The male then displays his anal fin in such a way as to bring his ocelli into view. The female attempts to pick up the spots, presumably mistaking them for ova. Whilst she is close to his anal fin, the male discharges sperm which are drawn into the female's mouth, fertilizing the eggs.

What are the physical and biotic factors contributing to the profuse speciation of *Haplochromis* in Lake Victoria?

In answering the question, we should consider the following factors:

(a)   Mouth brooding.

(b)   Selective mating for colouration and behaviour.

(c)   Choice of habitat and trophic specialization.

(d)   The possible seclusion afforded by the deeply indented bays of the lake.

(e)   The fact that cichlids do not migrate through rivers connecting lakes.

(f)   The small separate lakes that existed within the Lake Victoria basin in the early and later stages of lake formation.

Factors (a) and (b) would *strengthen* interspecific barriers, but are not thought likely to have initiated speciation. Factor (c), by itself, appears unlikely to be of primary importance in speciation because of the large amount of overlap in trophic and habitat requirements between different species. However, as we have pointed out earlier, trophic and habitat specializations can enormously speed up the morphological and genetic divergence of isolated populations (see also Section 6.1). Factor (d) is a possible contributing factor, but the evidence does not support it strongly: there is little apparent geographic differentiation morphologically, and no species is restricted in its geographical distribution within the lake. (Of course, this does not preclude the possibility that they were restricted in the past, and that genetically they may be differentiated into geographic races within species, even though morphologically this is not evident.) This leaves (e) and (f), which appear to be the most significant factors from the geological history of the lake. Primary isolation was possible in the small lakes, and absence of migration through rivers would reinforce this separation.

As already said, the lake started off as several lakes, and these were joined together and then separated several times as water levels rose and fell. P. H. Greenwood envisages that each time the populations were split up when the lake divided, the number of species increased in these small bodies of water. While isolated, the populations would undergo reorganization and divergence as a result of random genetic drift, founder effects and adaptive ecological shifts. Moreover, the peculiar courtship and mouth brooding behaviour would tend to reinforce separation and to maintain it when isolated populations became sympatric again. Thus when different bodies of water coalesced, the species originally present would have given rise to possibly as many species as there had been separate bodies of water. In the next round of fragmentation, the number of species would increase again. This would lead to a phenomenal rate of speciation.

> **ITQ 10**   It is estimated that there are now about 200 species of *Haplochromis* in Lake Victoria. How many rounds of fragmentation and coalescence would have been necessary to produce this number, if there was a single ancestral species? Assume that, on average, six separate bodies of water formed at each fragmentation, and each species gave rise to one other species in each small lake formed.

Concerning feeding habits, Greenwood considers that the range of anatomical and ecological diversity in the cichlid fishes far surpasses that found in either the Galapagos finches or the Hawaiian honeycreepers. He believes that this may have come about because a fish skull is a better starting point for functional modifications than is a bird skull.

The primary factor encouraging adaptive shift may well have been the lack of competition in the new habitats. It is noticeable that in all the '*adaptive trends*' within a *Haplochromis* lineage, presumed intermediate forms in the lineage are still represented by living species. This has important implications. The so-called trends in the fossil record, like those found in *Haplochromis*, are not necessarily true lineages; the 'intermediates' may not have been intermediate stages at all, but could each have been derived independently from some ancestor (often unknown). They may not form part of the actual pathway by which some very advanced form evolved from a more primitive one.

6 Sympatric speciation

Despite there being numerous cases of speciation in which geographic isolation appears to have played an indispensible role, there are circumstances in which sympatric speciation can, in theory, arise within the continuous range of distribution of a species, that is, without geographic isolation. Arguments about the relative importance of allopatric versus sympatric speciation are often semantic. For instance, in the proposed evolution of *Clarkia* species (see Section 4.1.2), a phase of drastic reduction in population density would in principle lead to geographically isolated groups even though these occurred within the once continuous range of the species. The question that should properly be asked is whether speciation could occur within a continuous range of distribution of a species, without the intervention of external barriers. In other words, could speciation occur through behavioural, physiological or genetic changes in the organisms themselves? For example it is conceivable that, in organisms with low vagility, local inbreeding groups may form and particular chromosome rearrangements become fixed (see Section 4.1.3). The reduction of fertility in structural heterozygotes would be a partial barrier which could encourage genetic divergence and speciation, particularly if augmented by an ethological mating barrier.

True sympatric speciation is conceivable in the extreme form of habitat isolation seen in insects with specific host plants. A divergence in the choice of host plant on which to feed or lay eggs could initiate strong habitat isolation leading to speciation, particularly if reinforced by a tendency of adults to breed and lay eggs on the foodplant on which they were reared. For example, in the sawfly *Pontania salicis* a number of races exist, and the larvae of each race produce galls on different species of willow. A race normally forming galls on *Salix andersoniana* was transferred to *S. rubra*. In the first generation the majority of the larvae died, but there was less heavy mortality in later generations. After four generations, three further generations were raised in which a choice of willows was provided. It was found that the sawflies continued to lay eggs on *S. rubra*, the new host plant.

Sympatric speciation is also likely to occur as a result of hybridization followed by polyploidy (i.e. allopolyploidy) in higher plants.

> ITQ 11 (a) Why are hybrids frequently sterile?
>
> (b) How could polyploidy overcome sterility in hybrids?
>
> (c) What factors would contribute to the isolation of the new polyploid species from the parental species?

The extensive involvement of polyploidy (both allopolyploidy and autopolyploidy) in the speciation of higher plants is indicated by the following. G. L. Stebbins estimates that 30–35 per cent of all species of flowering plants are polyploid and have gametic chromosome numbers that are multiples of the 'basic' number for their genus. However, other species, with higher chromosome numbers that are not multiples, may be secondary derivatives of polyploids. V. Grant, working on the assumption that species with numbers higher than $n = 13$ are polyploids, estimates that 47 per cent of all angiosperms are polyploids. This compares with estimates (on the same basis) of 1.5 per cent for conifers and 95 per cent for ferns and similar plants.

In many cases it has been possible to reconstruct the sequence of the hybridization and polyploidization events involved in the formation of a new species. Thus the common hemp-nettle (the tetraploid *Galeopsis tetrahit* in which $2n = 32$) grows throughout the Palearctic region. A. Muntzing was able to 'synthesize' it by crossing the diploid species *G. pubescens* and *G. speciosa* (both with $2n = 16$). The diploid hybrid had irregular meiosis and was fairly sterile. Eventually, however, reproduction occurred and a second generation (F_2) was obtained that included one triploid individual. This must have arisen from the fertilization of a diploid gamete by a haploid one. On backcrossing the triploid to *G. pubescens*, a tetraploid plant resembling *G. tetrahit* was obtained. This 'artificial' *G. tetrahit* was the result of two accidents in meiosis.

6.1 Sympatric speciation and disruptive selection

Thoday is the chief champion of the view that disruptive selection, or selection for distinctly different phenotypes within a population (see Unit 10), could give rise to sympatric speciation. Tauber and Tauber (1977) describe a case of sympatric speciation which they believe conforms to a model based on disruptive selection. The sibling species of lacewings, *Chrysopa carnea* and *C. downesi* (Neuroptera), occur sympatrically in northeastern United States. In nature, the two species are reproductively isolated through differences in habitats and seasons of reproduction. Under laboratory conditions, they hybridize freely and produce fully viable F_1 and F_2 offspring. *Chrysopa carnea* produces several generations between late spring and summer each year, and inhabits grassy areas and meadows. During the reproductive period, the adults are pale green in colour and match the light green foliage. At the end of summer, the adults enter diapause (a resting stage), turn from pale green to reddish brown, and move to the senescent foliage of deciduous trees. In contrast, *C. downesi* breeds once a year in early spring and is restricted to conifers throughout the year. The very dark green colour of *C. downesi* adults camouflages them in their habitat among the conifers during both reproduction and diapause.

In hybridization experiments, it was found that *C. downesi* was homozygous for the semi-dominant allele G_2 which produced the dark green colour, whereas *C. carnea* was homozygous for G_1, the allele giving the light green phenotype. The heterozygous $G_1 G_2$ was intermediate in colour. In addition, two other loci were involved in the control of the breeding cycle. *Chrysopa downesi* was homozygous recessive for both these loci, i.e. its genotype was *aa, bb*. The authors suggest the following sequence of events in the derivation of *C. downesi* from a *C. carnea*-like ancestor:

1 Disruptive selection favoured homozygotes in two different habitats—$G_1 G_1$ in grassy meadows and $G_2 G_2$ on conifers. The heterozygote, $G_1 G_2$, was selected against because of its reduced protective coloration in both habitats.

2 Habitat separation led to assortative mating, even though actual mating was random.

3 Selection for reproduction in early spring in *C. downesi* possibly resulted from an abundance of arthropod predators in summer.

4 Reproductive isolation was thus complete between *C. downesi* and *C. carnea*.

6.2 Summary of Sections 5 and 6

To conclude our study of speciation, here is a summary of the factors contributing to the formation of new species.

1 *Extrinsic factors*

(a) Isolation involving geographical or ecological barriers.

(b) Formation of small isolates by catastrophic reduction in populations due to selection by climatic factors or by infectious diseases.

(c) Availability of niches, and the relaxation of natural selection.

2 *Intrinsic factors*

(a) Chromosomal repatterning and reproductive isolation due to reduced fertility of heterozygotes.

(b) Genetic drift and founder effects in small isolated populations.

(c) Social structuring, favouring the formation of local subpopulations.

(d) Reduced vagility or dispersibility of local partial isolates.

(e) Propensity for migration, leading to the colonization of niches outside the main range of distribution.

(f) Disruptive selection due to spatial and/or temporal heterogeneity of the environment.

(g) Mating preferences.

(h) Habitat preferences.

With the possible exception of geographical isolation, there is no consensus as to the relative importance of these factors. Cases can be found in which any one of these factors appears to play a more or less critical role.

It is significant, however, that the isolation of one population from another often initiates large evolutionary changes—the sort that lead not only to speciation but to great morphological changes. This calls into question the role of natural selection in evolution. Is natural selection the 'creative force' which initiates new adaptations by a positive selection for small advantageous mutations, or is it merely the 'editor', a primarily conservative force eliminating the unfit by 'negative' selection? The 'latter view is adopted increasingly both by the critics of neo-Darwinism and by some professed neo-Darwinists themselves. They believe that major evolutionary change arises only when natural selection or competition is relaxed. This topic is considered further in Unit 13.

Finally, what light can palaeontology throw on modes of speciation?

☐ What are the two contrasting models of speciation proposed by palaeontologists?

■ You should recall from Section 3 of Unit 8 that traditionally speciation has been thought of as a gradual process. This has been challenged more recently by the 'punctuated equilibrium' model.

The gradualist model assumes that whole fossil populations have undergone gradual and continuous changes over long periods of time, so that new species have arisen by the gradual transformation of older species. A few examples of such gradual species transformations in the fossil record have been recorded. A celebrated case is that of the Late Cretaceous sea-urchin, *Micraster*, found in the chalk of northwestern Europe. In one of the principal lines of descent, changes in the shape of an ancestral species, *Micraster leskei*, have given rise to a new species, *Micraster cortestudinarium*, over a period of approximately 5 Ma. These changes in shape have been interpreted as adaptations to deeper burrowing into the sea floor by the descendent species—*Micraster leskei* appears only to have burrowed just beneath the surface. However, such records of gradualism are very rare despite extensive and diligent searching by generations of palaeontologists. The traditional explanation is that the fossil record is so incomplete that we can only ever see small segments of phylogenies. A major problem with this argument is that it would seem to be necessary to suppose that the same segments of phylogenies had been preserved in widely separated areas, for the science of stratigraphy is founded on the correlation in time of different sequences on the basis of their containing corresponding fossil species. This correspondence implies an unlikely conformity in local depositional events.

The alternative view—the punctuated equilibrium model—considers the 'gaps' as reflecting reality. According to this model, speciation usually involves rapid changes in small, localized populations. The resulting new species then remain in a state of evolutionary 'stasis', or at least show only trivial microevolution (such as increase in size), throughout the rest of their existence. Since the small initiating populations may not have been preserved, or may not yet have been located even if they have been preserved, most species can be expected to make sudden 'appearances' in the fossil record.

☐ What conclusion was reached in Unit 8 concerning the relative importance of the two modes of speciation in macroevolution?

■ The contrast between chronospecies rates and group rates of evolution, and the occurrence of living fossils, implies that punctuational speciation may play the greatest role in macroevolution.

The fossil record thus seems to indicate that most speciations involve rapid change in relatively small populations. However, 'rapid' to the palaeontologist may encompass a time-scale of tens or even hundreds of thousands of years. Clearly such a time-scale can include anything from a rapid but essentially continuous genetic change such as that advocated by Mayr in his concept of the 'genetic revolution', to sharply discontinuous genetic events such as the 'systemic mutations' advocated by Goldschmidt. Perhaps the most important point is that the record appears to support the importance in speciation of localized small populations—a feature of many of the biological models of speciation discussed in this Section.

45

6.3 Objectives and SAQs for Sections 5 and 6

Now that you have completed these Sections, you should be able to:

(a) State the elements of the model of allopatric speciation and contrast the views of Mayr and Dobzhansky about the origin of reproductive isolation.

(b) State the 'intrinsic' and 'extrinsic' factors influencing the effectiveness of geographic isolation.

(c) Contrast the allopatric and sympatric theories on the origin of reproductive isolation mechanisms.

(d) Evaluate the significance of the different factors involved in particular cases of speciation.

(e) Distinguish between and give examples of allopatric and sympatric speciation.

To test your understanding of these Sections and of the television programme 'The picture wings of Hawaii', try the following SAQs.

SAQ 7 (*Objectives a–e*) Assess the importance of the following factors in the profuse speciation of Hawaiian drosophilids:

(a) geographical barriers;

(b) founder effect;

(c) habitat preference;

(d) sexual selection.

SAQ 8 (*Objectives c and e*) Devise suitable experimental tests for stages 1 and 3 of the sequence of speciation of *Chrysopa downesi* from *C. carnea* proposed at the end of Section 6.1.

General references

LAMARCK, J. B. (1809) quoted by OSBORN, H. F. (1929) *From the Greeks to Darwin*, Charles Schribner.

MAYR, E. (1957) Species concepts and definitions, in *The Species Problem*, Publication 50, AAAS, Washington.

PLATE, L. (1914) quoted by MAYR, E. (1957).

TAUBER, C. A. and TAUBER, M. J. (1977) Sympatric speciation based on allelic changes at three loci: evidence from natural populations in two habitats, *Science*, **197**, 1298–1299.

Further reading

DOBZHANSKY, T., AYALA, F. J., STEBBINS, G. L. and VALENTINE, J. W. (1977) *Evolution*, W. H. Freeman.

LEWONTIN, R. C. (1974) *The Genetic Basis of Evolutionary Change*, Columbia University Press

MAYR, E. (1970) *Populations, Species and Evolution*, The Belknap Press of Harvard University.

FUTUYAMA, D. J. (1979) *Evolutionary Biology*, Sinauer Ass. Sunderland. Mass.

GREENWOOD, P. H. and FOREY, P. F. (eds) (1981) *Chance, Change and Challenge: The Evolution of the Biosphere*, Cambridge University Press/British Museum of Natural History.

WHITE (1978) *Modes of Speciation*, W. H. Freeman.

Acknowledgements

Grateful acknowledgement is made to the following for permission to reproduce material in this Unit:

Figure 5 courtesy of Dr J. D. A. Delhanty; *Figure 12* from Wilson *et al.* (1974) 'The importance of gene re-arrangement in evolution' in *Proc. Nat. Acad. Science*, vol. 71; *Figure 13* from King and Wilson (1975) 'Evolution at two levels in humans and chimpanzees' in *Science*, vol. 188. Copyright © 1975 by the American Association for the Advancement of Science; *Figures 15 and 16* from Greenwood, P. H. (1974) *Cichlid Fishes of Lake Victoria, East Africa*, British Museum

ITQ answers and comments

ITQ 1 Yes, most of the different breeds actually interbreed, and all can potentially do so.

ITQ 2 The identity of alleles, I, over the two loci is

$$I = \frac{J_{xy}}{\sqrt{J_x J_y}}$$

$$= \frac{0.25}{\sqrt{0.57 \times 0.70}}$$

$$= 0.40$$

The genetic distance D is

$$D = -2.3 \log_{10} I \qquad\qquad D = -\ln I$$

$$= -2.3 \log_{10} 0.40 \quad \text{or} \quad = -\ln 0.40$$

$$= 0.92 \qquad\qquad = 0.92$$

(Logarithms can be found either by looking them up in a suitable Table or by using a calculator which has a log function.)

ITQ 3 In disturbed areas, the usual habitats of each species disappear so that isolation breaks down and the species meet more often. Hybridization thus occurs more frequently.

ITQ 4 Recombination occurs at meiosis (see Table A1). In a hybrid between species, recombination between the parental genotypes will occur during meiosis in the first generation hybrid.

ITQ 5 The results indicate that altered butterflies, i.e. non-mimics, are probably subject to predation by birds whereas mimics are not.

ITQ 6 (a) Proteins are coded by specific base sequences in DNA. A change in the base sequence of DNA may produce a change in the amino acid sequence of the protein. If this involves an alteration in net charge, the mutant protein will have a different electrophoretic mobility from the normal protein. So electrophoretic variation can provide an estimate of underlying genetic variation.

(b) The limitations are that it invariably leads to an underestimate of genetic variation for the following reasons:

(i) Only amino acid changes involving an alteration in net charge are likely to be identified, i.e. about one-third of the amino acid substitutions. Refinements of electrophoretic techniques do, however, allow more substitutions to be distinguished.

(ii) Synonymous codon changes in the DNA, i.e. those which do not result in an altered amino acid, will not be recognized. Remember from the Foundation Course that more than one codon codes for most of the amino acids. For example, the amino acid leucine is coded for by six different codons.

(iii) Genes without a recognizable (or easily isolated) protein as a gene product cannot be studied electrophoretically.

ITQ 7 *D. insularis* and *D. pavlovskiana* are the most widely separated sibling species. The separation between the sibling pairs *D. equinoxialis*–*D. paulistorum*, *D. equinoxialis*–*D. tropicalis*, *D. equinoxialis*–*D. willistoni*, *D. paulistorum*–*D. willistoni*, and *D. paulistorum*–*D. tropicalis*, are all greater than that between the non-sibling pair *D. pavlovskiana*–*D. paulistorum*.

ITQ 8 Mutations very rarely occur, and most mutations are deleterious. Advantageous mutations must be orders of magnitude less frequent than deleterious ones, so a mutation having a large phenotypic effect which is at the same time advantageous is almost inconceivable. Large changes in phenotype involve changes in coordinated characters and functions, and it is difficult to conceive of a random mutation which would not disturb this coordination. Moreover, such 'hopeful monsters' have never been observed. Arguments of this kind have been used against both systemic mutations and sudden changes (saltations), from Darwin's day to the present. We return to this problem in Unit 13.

ITQ 9 (a) About 130;

(b) about 10. These numbers are very approximate because the curves represented by the histograms are not smooth.

ITQ 10 Three rounds of speciation by fragmentation would produce $6 \times 6 \times 6 = 216$ species.

ITQ 11 (a) Sterility in hybrids is mainly caused by the lack of homology between the chromosomes of the two parental species. This means that normal pairing cannot occur in meiosis and viable gametes will not be formed.

(b) In an allopolyploid, *all* the chromosomes become duplicated; pairing can occur and meiosis proceeds to give viable gametes. Thus in a hybrid with chromosome sets (m) and (n), polyploidy produces a diploid karyotype $(2m + 2n)$ and meiosis is normal.

(c) Let us represent the chromosome complements of the parental species by $(2m)$ and $(2n)$ respectively. The F_1 hybrid and the polyploid are then $(m + n)$ and $(2m + 2n)$ respectively. On back-crossing the polyploid to the parental species, say with $(2m)$ chromosomes, the resulting progeny is a triploid which has a chromosome complement of $(2m + n)$, i.e. $(m + n)$ from one parent and (m) from the other. This triploid is sterile, again because of a disturbance to meiosis—the single set of chromosomes, e.g. (n), cannot pair. Thus the polyploid is reproductively isolated from both of the parental species.

SAQs answers and comments

SAQ 1 (a) Asexual species by definition do not reproduce by crossing; each is in effect a clone or a collection of clones (a clone is the genetically identical progeny of an individual). Therefore they continue to breed true, whether other species are there or not.

(b) Palaeontological species are recognized solely on morphological grounds, although stratigraphic separation between two forms is often taken to indicate distinctiveness of species. For species with the same stratigraphic distribution, there is very little to warrant the assumption of reproductive isolation.

(c) Allopatric species are by definition geographically isolated and hence prevented from interbreeding. Tests could be performed in the laboratory to see whether the species can successfully interbreed, but their relevance to whether the species would interbreed in nature if the species became sympatric is questionable.

(d) By definition, the two species ought to be reclassified as belonging to a single species.

SAQ 2 The data may be summarized as follows:

1 *D. heteroneura* females discriminate absolutely against *D. silvestris* males.

2 *D. silvestris* females discriminate strongly against *D. heteroneura* males.

3 F_1 hybrid females discriminate strongly against *D. silvestris* males.

4 There is no evidence of sterility or hybrid breakdown when mating is successful.

The conclusion is that the reproductive barriers between the two species are all of the pre-mating ethological (behavioural) type based on female choice.

SAQ 3 Factor (i) could not have been very important in race formation in *H. erato* and *H. melpomene*, because precisely the same genes for pattern would have had to be fixed by drift or founder effect in the geographic isolates of both species. Factor (ii) also could not have been important, as it cannot explain how the patterns in the races of both species came to resemble each other. Factors (iii) and (iv) could explain the convergence in pattern as being an adaptation to the same physical and biotic environment. Factor (v) would contribute to increased divergence between isolates of the same species.

SAQ 4 The completed dendrogram is shown in Figure 17.

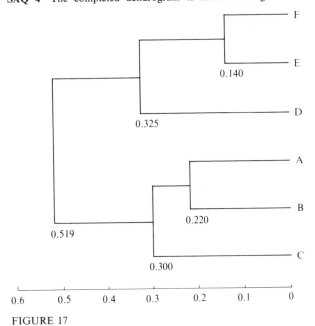

FIGURE 17

The procedure is as follows:

(i) Identify the two most closely related pairs of species—these are E–F and A–B. Connect these pairs on the dendrograms, at the appropriate genetic distances 0.140 and 0.220.

(ii) Identify the species most closely related to *both* E and F, i.e. D. This is connected to the pair E–F by a branch, representing the average genetic distance 0.325.

(iii) Identify the species most clearly related to both A and B, i.e. C, and connect as in (ii).

(iv) Finally connect the two clusters by a branch representing the average genetic distances between all possible species pairs between E, F and D on the one hand and A, B and C on the other.

SAQ 5 The phylogeny is:

$$4 \rightarrow 2 \rightarrow 1 \rightarrow 3$$

i.e. two inversions followed by a translocation.

This is based on the principle of parsimony, i.e. the minimum number of steps required, and on the information that the Moon Island species is possibly the oldest.

SAQ 6 (a) When the rate of chromosomal evolution, estimated from the average change in chromosomal number or arms per genera per million years, is plotted against the total rate of speciation (estimated from the sum of net speciation and average extinction rates), a highly significant correlation is obtained. This is *hypothesized* to indicate a causal relationship between chromosomal evolution and speciation, but the evidence required to *establish* such a causal relationship does not yet exist.

(b) The arguments are as follows:

(i) On anatomical and physiological grounds, humans and chimpanzees are classified in different genera.

(ii) DNA hybridization, however, indicates that there is 100 per cent homology between the DNA of human and chimpanzee.

(iii) Immunological comparisons of homologous proteins demonstrated that the mean difference in amino acid sequence is very small.

(iv) Frogs and mammals differ widely in their rates of organismal evolution (as estimated by the rate of development of an inability to form viable hybrids between species) though their proteins have evolved at the same rate; this shows that organismal evolution can be decoupled completely from genetic change.

(v) Similarly, protein (genetic) evolution has proceeded at the same rate in humans and chimpanzees, whereas organismal change has been much faster in humans.

(iv) It is postulated that the observed genetic change, which concerns only the structural genes, does not reflect the genetic changes that have taken place in the regulatory genes. These regulatory genes either do not have gene products (in terms of identifiable molecules), or their gene products have not yet been discovered.

SAQ 7 (a) Geographical barriers. The existence of many isolated islands as well as the subdivision of each island by erosion and lava flow (producing kipukas, for example) provide abundant opportunity for profuse allopatric speciation (see the television programmes 'The picture wings of Hawaii' and 'Islands within islands within islands').

(b) Founder effect. This is considered a major factor in determining the divergence of new colonizing populations of *Drosophila*. It is believed that successive colonization of islands by *single* females with fertilized eggs have been involved in the multiplication of species.

(c) Habitat preference. Many *Drosophila* species exhibit specialization in plant hosts (see the television programme 'The picture wings of Hawaii'). These species could have evolved following an initial choice of habitat by founding members.

(d) Sexual selection. In nature, sexual isolaton is very strong between some Hawaiian species (e.g. between *D. silvestris* and *D. heteroneura*), though under no-choice conditions in the laboratory, fertile hybrids can be obtained (Section 3.2.2). Some workers believe that selection for mating behaviour can by itself lead to speciation, as explained in the television programme 'The picture wings of Hawaii'. The possibility exists that the different mating behaviour could have evolved independently in geographically isolated populations which then become sympatric (Section 5.1).

SAQ 8 To test stage 1, the mortality due to predation for all these genotypes in both environments could be assessed; laboratory tests with suitable predators could also be carried out, so that mortality due to predation could be directly observed.

To test stage 3, ecological studies could be made to ascertain mortality due to predation at different times from early spring to late summer. The experimental design could involve marking a counted number of lacewings (of each genotype where appropriate) and recapturing them after a period of time in a previously cleared field.

UNIT 13 MACROEVOLUTION AND DEVELOPMENT

Contents

TABLE A1   List of terms and concepts assumed from the Science Foundation Courses*

Term	S100 Unit No.	S101 Unit No.	Term	S100 Unit No.	S101 Unit No.
amino acid	10, 14, 17	16/17, 23	molecular biology	17	25
chromatid	19	25, AV	molecular genetics	17	25
crossing over	19	19, 25	nucleotide	14, 19	24
DNA	13, 17	25	polypeptide	—	24
F_1 hybrid	—	19	protein synthesis	17	25
gene	17	19	RNA	17	25
genetic code	17	25	ribosome	14, 17	25
indices	MAFS†	MAFS†	ribosomal RNA (rRNA)	17	25
logarithm	MAFS	MAFS	transcription	17	25
meiosis	17, 19	25	transfer RNA (tRNA)	17	25
messenger RNA	17	25	translation	17	25
mitosis	17, 19	25			

* The Open University (1971) S100 *Science: A Foundation Course*, The Open University Press; or The Open University (1979) S101 *Science: A Foundation Course*, The Open University Press.

† The Open University (1970/1979) *Mathematics for the Foundation Course in Science*, The Open University Press.

TABLE A2   List of terms and concepts from previous Units and/or developed in this Unit or in the Glossary‡

Term	Section No. in this Unit	Term	Section No. in this Unit	Term	Section No. in this Unit
*acceleration	8, 8.2	genetics of pattern	7.1	polygenes (10)	7.1
*allometric equation	6.3, 8, 8.1	*genome‡ size	4, 4.1	polyploidy‡ (12)	4, 4.1
*allometry (positive and negative)‡	6.3	geographical race (12)	7.1	pre-adaptation (1, 4)	3.1, 3.2
allopatric speciation‡ (12)†	2, 3.1	*Gryphaea* (development in)	8.1	pre-adaptive phase	3.1, 3–3.2
archetypes‡ (3, 4, 6)	6	Haeckel's law of recapitulation	5	producer genes, *P*	4.2
Baldwin–Morgan effect	9.2	*heterochrony‡	6.3, 8–8.3	*progenesis‡	8.2, 8.3, 9.1.2
bone development (5)	9.1.1	*hypermorphosis	8.2	*proportioned dwarfism or gigantism	8.2
*Britten–Davidson model	4.2	*inheritance of acquired characters	9.2–9.4	punctuated equilibrium‡ (8, 12)	2
*canalization	9.3	integrator genes, *I*	4.2	*quantum evolution	3, 3.1
chronospecies (1, 8)	2	*interspecific allometry	6.3	*r*- and *K*-selection‡ (11)	8.3
*clock model of heterochrony	8.2	*intraspecific allometry	6.3	*Ranunculus* (water crowfoots)	9.1.3
coefficient of geometric similarity, *s*	8, 8.1	*isometry	6.3, 8	reassociation of DNA (12)	4.1
*coincident variation	9.2	*macroevolution‡ (1, 8)	1, 2, 4.1, 8.3	*recapitulation‡	8–8.3
cytoplasmic inheritance	7.1	Mayr's model of allopatric speciation (12)	3.1	receptor genes, *R*	4.2
denaturation of DNA (12)	4.1	metamorphosis‡ in insects	9.1 9.1.2	*regulatory genes‡ (12)	4.2
*developmental adaptability	9.1, 9.3	*microevolution‡ (1)	2	*repetitive DNA‡	4.1
*developmental homeostasis‡	9, 9.3	modifier genes	9.3	*retardation	8, 8.2, 8.3
*developmental plasticity	9.1, 9.3	neo-Darwinism (1)	1, 1.1, 3.2	*saltatory evolution	2, 3, 3.2, 7
electrophoresis (10, 12)	2	*neoteny (1, H.E.§)	8.2, 8.3, 9.1.2	sensor genes, *S*	4.2
*epigenetic landscape	9.3	*neutralist theory (10)	1.1	Simpson's model (of quantum evolution)	3.1
eukaryote‡ (2)	4	*ontogenetic allometry	6.3	somatic development	8, 8.2
founder effect (12)	2	ontogeny‡	4.2, 5, 8	synthetic theory of evolution	1
*gene duplication	4.1	organic selection	9.2	titanotheres	6.3
*genetic assimilation‡	9.3	orthogenesis	6.3	transformation	6.2
genetic drift‡ (10, 12)	1, 2	*paedomorphosis‡ (H.E.)	8–8.3, 9.1.2	*Von Baer's laws of form and development	5, 6
genetic revolution (12)	2, 3.1	*phenocopies	9.3	variation of weight with size (1)	6.1
		*phenotypic plasticity	9, 9.1.3		
		*phylogenetic allometry	6.3		
		phylogeny‡ (1, 8)	5, 8		

* The most important terms are indicated by an asterisk.

† The number of the previous Unit is given in parentheses, after the term.

‡ Items followed by a double dagger are to be found in the Glossary (Section 4 in the Handbook).

§ Home Experiment Notes.

Objectives

When you have completed this Unit you should be able to:

1 Define and recognize the best definitions of the terms, concepts and principles marked with an asterisk in Table A2 and distinguish between true and false statements concerning these terms. (ITQs 1, 14 and 15; SAQs 1, 3, 7, 9 and 10)

2 State the major difficulties involved in accounting for macroevolution in terms of microevolutionary processes and the models advanced to explain it. (SAQ 1)

3 Describe and interpret data relating to the model of gene regulation based on repetitive DNA. (ITQ 4; SAQs 2 and 3)

4 Contrast Haeckel's doctrine of recapitulation with Von Baer's laws of development, and state their evolutionary significance. (SAQs 4 and 5)

5 Use the allometric equation and distinguish between different allometric relationships. (ITQ 11; SAQ 6)

6 Recognize data suggesting the existence of genetic controls in the timing and the rates of developmental processes. (SAQ 8)

7 Draw inferences concerning processes involved in heterochrony from allometric data. (ITQs 10 and 11; SAQs 7 and 9)

8 Draw and interpret clock models of heterochrony and state the morphological consequences of different heterochronic processes. (SAQs 7 and 9)

9 State, in general terms, the relationship between genes and phenotypes. (SAQ 11)

10 Evaluate, criticize and suggest experiments on the mechanisms of canalization and the genetic assimilation of novel phenotypes. (ITQs 17 and 18; SAQs 10, 11 and 12)

Study Guide for Unit 13

In Units 10 to 12, evolution was explained in terms of the natural selection of random mutations or of other genetic mechanisms such as recombination, gene flow and genetic drift. All of these explanations are contained within neo-Darwinism and its contemporary version, often called the synthetic theory of evolution (Section 1). Unit 13 is in many ways a departure from these theories. This is not to say that neo-Darwinism is 'wrong' but rather that there are other ways of looking at evolution which emphasize different mechanisms.

There is currently a move to 'bring back the organism' into evolutionary theory with a clear shift in emphasis from genes to form, development, and organization of living things. Whether this offers a more satisfactory explanation of evolution you will have to judge for yourselves. A list of further reading is included at the end of the Unit to enable you to find out more about these alternative views. A useful way to look at this Unit, regardless of where you stand in relation to neo-Darwinism, is to treat the ideas introduced here as possible additional mechanisms to those proposed by the neo-Darwinists. The relative importance of all these mechanisms is being actively debated. Evolution is nothing if not complex, and hence most likely to involve many causes. Any attempt to explain evolution in terms of a simple mechanism, whatever it may be, will certainly not succeed.

Try to read straight through the Unit but, if very short of time, you could omit Section 4 (although you will not be able to achieve all Unit Objectives if you do). This Section is relatively self-contained and it requires some background knowledge of molecular biology from the Science Foundation Courses (see Table A1). Sections 6, 7 and 8 should be studied with particular care because they describe important concepts: you will need a calculator and/or log tables for Section 8.

The television programme 'Time for change' is particularly related to Sections 5, 6, 8 and 9 of this Unit and is best seen after reading these Sections.

1 Introduction

In the first part of the Course, we painted a broad canvas of evolution, starting from the origin of life and leading on to the diversification of the major phyla and classes of organisms. This was followed by an overview of patterns and rates of evolution, and of the physical setting in which the whole evolutionary 'drama' was enacted.

Beginning with Unit 10, we turned from a predominantly descriptive to an analytic mood, and attempted to deal with the *mechanisms* responsible for evolution. The dominant theory today is, of course, neo-Darwinism, or its contemporary version, the synthetic theory. This explains evolution in terms of the natural selection of random mutations, supplemented by other genetic mechanisms such as recombination, gene flow and genetic drift. The subject matter of Units 10–12 essentially covers most aspects of the synthetic theory. To many, though by no means all, neo-Darwinists this theory is necessary and sufficient to account for the whole of evolution. Thus Maynard-Smith (1969), in defending the theory from its not-too-infrequent critics, stated:

synthetic theory of evolution

> So far I have been describing a set of properties of organisms (multiplication, heredity and variation) or more precisely, a set of properties which neo-Darwinism assumes all organisms to have. This is not by itself a theory of evolution. The theory of neo-Darwinism states that these properties are *necessary and sufficient* to account for the evolution of life on this planet to-date. [italics ours]

This is a rigorous statement of the theory of neo-Darwinism which makes explicit an assumption that nearly all neo-Darwinists tend unconsciously to adopt—though they may not support it openly when challenged. In Unit 13, we take issue with this assumption by presenting other mechanisms which we believe to be involved in evolution. Though individual neo-Darwinists may recognize the importance of some of these additional mechanisms, it must however be realized that they are not part of the theory of neo-Darwinism. Standard textbooks on evolution give a detailed exposition of natural selection and other elements of the synthetic theory. Very few such accounts mention development of organisms, and none deal with the evolutionary implications of mechanisms of development as we do here.

Both Darwinism and neo-Darwinism have been criticized ever since they were invented. One major weakness of these theories is their failure to explain macroevolutionary phenomena, i.e. the origin of new types or forms of organisms. Neo-Darwinism tells us a great deal about how moths change in adaptive coloration but almost nothing about how there came to be moths in the first place, for neo-Darwinism does not address the latter type of problem at all. We do this to some extent in Unit 13.

1.1 The neutrality of molecular evolution

In Unit 12 we examined at length the nature of intra- and interspecific variation and the circumstances favouring the formation of new species. We found that organismic evolution appears to be decoupled from genetic or molecular evolution. Furthermore, the most rapid evolution seems to take place in small, isolated populations where competition is relaxed. The differences between molecular and morphological rates of evolution have already been pointed out in Unit 8 (Section 4.3). Morphological and taxonomic rates vary greatly through geological time, whereas molecular evolution seems to occur at approximately constant rates.

The most striking contrast between morphological and molecular evolution is seen in the so-called living fossils such as *Latimeria*, the coelocanth. Their proteins have undergone just as many amino acid substitutions as those in the groups evolving fastest morphologically. These observations are the strongest evidence in favour of the *neutralist* theory of molecular evolution, which maintains that most successful amino acid substitutions in proteins have been due to selectively neutral mutations—those that are neither advantageous nor deleterious—becoming fixed by genetic drift in finite populations. This is contrary to the more orthodox (or *selectionist*) view which maintains that amino acid substitution occurs by the natural selection of advantageous mutations. Both neutralists and selectionists agree that the majority of mutations are deleterious and are eliminated from the

neutralist theory

population sooner or later, contributing nothing to molecular evolution. The question, therefore, is whether the substitutions that do occur are due to random genetic drift or to natural selection.

One virtue of the neutralist theory is that it makes a specific prediction: that amino acid substitution occurs at constant rates. Since neutral mutations arise at a constant rate and are substituted (or fixed) at this rate, it follows that *the rate of substitution of neutral alleles is constant, independent of population size and any other parameter.*

The approximate constancy of molecular rates of evolution, as we have seen, conforms quite well to the prediction from the neutralist theory, and contrasts with the varying tempo of morphological evolution. This apparent decoupling of genetic from organismic evolution raises the whole question of the relationship between genes and the development of form. Without a clear understanding of this relationship it is questionable whether a purely genetic theory, such as neo-Darwinism, can adequately explain evolution.

In the first part of this Unit we shall look at current attempts to restructure the genetic theory. Later, we shall develop the thesis that the real problem of evolution is how *forms*, or organisms, change in time. The neo-Darwinian theory is strictly a theory of genes, whereas the phenomenon to be explained is that of the transmutation of form. In the absence of a theory of how forms are generated and how genes can affect the development of form, neo-Darwinian theory offers at best an incomplete explanation of evolution.

2 Microevolution and macroevolution

Microevolution refers to phenomena which take place within relatively short time-scales. They include changes in allele frequencies in natural populations as the result of natural selection, migration, genetic drift, and other processes amenable to experimental studies and/or field observations. The evolutionary changes involved, if any, are small, and often due to the substitution of alleles in single genes. Units 10 and 11 dealt largely with these microevolutionary phenomena.

microevolution

Macroevolution, on the other hand, refers to phenomena which take place over geological time-scales—millions or tens of millions of years—and which are especially associated with the origination of major taxa. Large qualitative changes are often involved and novel features or complexes of novel features appear. (Different, though complementary, definitions of micro- and macroevolution were given in Unit 1, Section 2.6.)

macroevolution

> **ITQ 1** Identify the following phenomena as either micro- or macroevolutionary:
>
> (i) changes in frequency of black peppered moths in parts of the British Isles;
>
> (ii) adaptive radiation of the molluscs;
>
> (iii) origin of the eukaryotic cell;
>
> (iv) evolution of heavy metal tolerance in grasses;
>
> (v) evolution of the tetrapod limb.

Opinion is divided as to whether there is any real distinction between microevolution and macroevolution. The more orthodox neo-Darwinian view is that there is no difference at all; microevolutionary processes operating over immense time will give rise to macroevolution. In other words, the large changes observed are due to an accumulation of small changes over millions of years. The crux of the matter is whether *intermediate* stages of macroevolutionary changes have ever existed. What do we mean by intermediates? Obviously, if one accepts that new organisms arise by evolution rather than by special creation, new types do not turn up miraculously from nowhere. The question is how *small* the intermediate steps have to be. At one extreme, there are those who, like Darwin himself, believe that continuous variation is the stuff of evolution, and therefore that a continuous series of intermediates existed. At the other extreme, there are those who believe that the variations responsible for macroevolutionary changes were large discontinuous ones. Intermediates, even if they existed, consisted of discrete jumps. Certainly, no intermediate stages showing small continuous changes between organisms belonging to different phyla have ever been found, either in the fossil record or among living representatives. Intermediate jumps of varying sizes are apparent in the fossil record of evolution *within* phyla, for example in the radiation of molluscs described in Unit 4, but do not appear between the various metazoan phyla.

Traditionally, the absence of intermediates among living representatives was attributed to the extinction of intermediate forms due to competitive exclusion. The absence of extinct intermediates, in turn, was explained by the 'incompleteness of the fossil record'. The latter obviously applies to the origin of metazoan phyla which took place in Precambrian times. As opposed to this view, there have always been those who regarded the abrupt transitions between major taxa as real, and hence considered that macroevolution involves processes distinct from those in microevolution. In fact, careful stratigraphic analysis in recent years gives some support to this very old idea.

> ITQ 2   Recall from Unit 8 evidence in support of the claim that micro-evolution and macroevolution may involve different processes.

The emerging picture of evolution is that of 'punctuated equilibria'—long periods of stasis in which little or no change occurs, punctuated by short intervals (geologically speaking) in which rapid change takes place. Another way of looking at this is that there are two kinds of evolutionary processes: very slow and gradual *phyletic* changes which transform a species into another species (chrono-species) without one species splitting off from the other; and rapid discontinuous changes in which one species splits off from another. Thus, macroevolutionary changes seem to be concentrated in short episodes of speciation. This is consistent with the currently widely accepted model of allopatric speciation (Unit 12). Rapid, if not large, discontinuous changes are specifically predicted in the model and the general term *saltatory evolution* is applied to such rapid changes. ·

saltatory evolution

> ITQ 3   Which are the genetic processes postulated to give rise to large changes in allopatric speciation?

The so-called genetic revolutions in small isolated populations which are supposed to produce saltatory evolution have received little support from electrophoretic studies on protein polymorphisms—enzyme variants within populations (Unit 10). The only explanation within the genetic framework is to postulate changes in regulatory genes (see Unit 12, Section 4.3). We shall say more about regulatory genes in Section 4. For the moment we shall concentrate on the more traditional attempts to explain macroevolution.

3   Quantum evolution versus saltatory evolution

To account for the morphological discontinuities that exist between the major groups of organisms, two models have been proposed: quantum evolution, proposed by G. G. Simpson, emphasizes (despite its name) gradualism and continuity; saltatory evolution explicitly allows for discontinuities or saltations.

3.1   Quantum evolution

We have seen in Unit 12 that Ernst Mayr's model of allopatric speciation allows for rapid changes, especially in small founding populations, due to genetic revolution or drastic reorganization of the genome. The new genotypes are subject to new selective forces in the environment, as a result of which rapid *but continuous* morphological changes can take place. Mayr's model is an elaboration of Simpson's quantum evolution which involves three phases:

Simpson's model

(i)   an inadaptive phase, during which the population loses its adaptive equilibrium with its environment;

(ii)   a pre-adaptive phase, during which the population moves towards a new equilibrium;

(iii)   an adaptive phase, during which a new equilibrium is attained.

These three phases correspond in Mayr's model to those of (i) genetic reorganization, (ii) formation of new genotypes with the potential for improvement by natural selection (pre-adapted genotypes), and (iii) establishment of morphological adaptation to the new environment.

Both Mayr and Simpson emphasize continuity, though it is by no means a necessary consequence of their models. In fact, as has been pointed out, Mayr's model specifically *allows* for saltations (see Unit 12, Section 6.2).

3.2 Saltatory evolution

A number of theories postulate that 'large' genetic changes give rise to major changes in phenotype, which are then subject to natural selection. The formation of new types of organisms are presumed to occur in jumps.

Schindewolf (1936) noted that the morphological discontinuities correspond to the observed gaps in the fossil record. He regarded the gaps as evidence of abrupt transitions from one major adaptive type to another, rather than being due to deficiencies in the record. He suggested that new morphological types are attributable to disturbances in *developmental processes*, which can produce abrupt and profound changes in adult organization. He argued that such changes would occasionally, by chance, produce a form that is basically pre-adapted to a new way of life and this basic type would then be 'shaped' or 'perfected' by selection. This view is practically identical to that of R. B. Goldschmidt (see Unit 12, Section 4), if for 'disturbances in developmental processes' we read 'genetic disturbances'. According to this view, transitional stages between one basic type and another never existed, since the fundamental determining features of the new type arose instantaneously during development. There are thus two phases in the origin of a new adaptive type:

(i)   the 'pre-adaptive' phase, usually quite short, during which new types of organism arise abruptly;

(ii)   a succeeding, usually longer, period of continuous, gradual shaping of the new type—the 'adaptive' phase.

Recently, theories have been put forward about the evolution of the genome in terms of its content and organization. As these aspects are not covered explicitly by the neo-Darwinian theory, they have been referred to as 'non-Darwinian evolution'.

4   Evolution of genome size

> **Study comment**   This Section requires familiarity with basic concepts of molecular genetics. You should re-read relevant material on molecular biology and molecular genetics from the Foundation Course (see Table A1) if you find this Section difficult. If you are *very* short of time you may leave out this Section, but you will then be unable to achieve all the Unit Objectives.

Broadly speaking, genome size (or the total amount of DNA in the cell nucleus) increases with the grade of organization of eukaryotes (Unit 7), though the wide range of genome sizes among closely related species obscures this trend. If the *minimum* DNA content for various grades of organization is plotted, we obtain the relationship shown in Figure 1.

genome size

☐ Why is the *minimum* DNA content plotted rather than the *mean* for each major group?

■ To estimate the size of the original haploid genome. As polyploidy may have occurred within each group, the size of the haploid genome is taken to be the minimum amount of DNA sufficient to specify an organism of any particular grade.

Data such as those in Figure 1 have prompted the suggestion that the increase in genome size is correlated with and somehow responsible for the increase in complexity of organisms in evolution. The difficulty with this suggestion is that

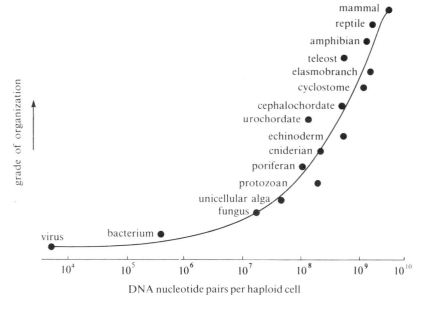

FIGURE 1 Minimum haploid DNA content (in nucleotide pairs per cell) in species at various grades of organization.

we still do not know how many genes are required to build even a bacterium, let alone a bee or a bat. The best known and understood genes are the structural genes which code for specific proteins, for ribosomal RNA and for transfer RNA molecules. Most of the known biosynthetic and catabolic pathways in higher organisms are already present in unicellular organisms. So, what exactly is the function of the extra DNA in higher organisms?

From Figure 1, it can readily be seen that the minimum DNA content of a bacterial cell is about 4×10^5 nucleotide pairs whereas that of a mammalian cell is about 3×10^9.

☐ Assume that the average polypeptide contains 300 amino acids. How many polypeptides can be coded by the minimum DNA in (i) a bacterium, and (ii) a mammal?

■ Since three nucleotides code for one amino acid, it takes 900 nucleotide pairs to code for one average polypeptide. The minimum DNA in bacteria is sufficient to code for $(4 \times 10^5)/900$ or 444 polypeptides, whereas the minimum DNA in mammals is sufficient to code for 3.33×10^6 polypeptides.

The number of identified gene products in all organisms studied is of the order of 10^3. This leads us to the seemingly inevitable conclusion that a large part of the DNA in eukaryotic cells may be redundant. Before we discuss this further, let us examine the basic characteristics of eukaryotic DNA.

4.1 Gene duplication

Some evolutionists believe that while allelic changes at existing gene loci may give rise to racial differentiation within species and to other microevolutionary phenomena, they cannot account for the large changes in evolution that are involved in the origin of major new types. It is claimed that such processes depend on the acquisition of new gene loci with new functions. Natural selection is seen primarily as a negative force which conserves existing functions but never initiates new ones, so that evolution can only occur by an 'escape' from natural selection. Such an escape, it is proposed, can be effected by *gene duplication*, in which a duplicated and therefore redundant copy of a gene accumulates mutations that give it, by chance, a new function. There is ample evidence for duplicated genes in the eukaryote genome, as shown by the investigations described below.

gene duplication

When eukaryotic DNA is sheared into small fragments of several hundred nucleotide pairs, denatured and then allowed to reassociate (see Unit 12, Section 4.2.2) the rate of reassociation for a particular sequence is directly proportional to the number of copies of that sequence present. Thus the rate of reassociation depends on both the complementarity of the nucleotide sequences and the concentration of the reacting sequences. If only single copies of sequences or genes existed, DNA from larger genomes would reassociate more slowly than that from smaller genomes. This is because for the same concentration, in terms of weight or of base pairs of DNA per unit volume, the solution of the smaller genome would contain more complementary pieces than that of the larger genome. In other words, for a given concentration of DNA, the smaller the genome the greater the chance that complementary pieces will meet up with one another and reassociate.

☐ Solutions containing 10^6 nucleotide pairs per litre are prepared from genomes A and B separately. These have 10^5 and 10^6 nucleotide pairs per genome respectively. Assuming that only single gene copies exist in both genomes,

    (a)   how many single complementary strands of each gene are there per litre, in the solutions of genome A and genome B respectively, and

    (b)   what are the relative rates of reassociation?

■ (a)   Ten in A and one in B.

    (b)   Reassociation for genome A is ten times faster than for B.

When carrying out experiments on the reassociation of denatured DNA from eukaryotes, Britten and Kohne (1968) observed that the DNA of higher organisms contains two fractions which reassociate at vastly different rates: a main fraction which reassociates relatively slowly, and a smaller fraction which reassociates very much more quickly. For example, calf thymus DNA has a fraction comprising about 38 per cent of the total DNA which reassociates at about 10^5 times the rate of the bulk fraction. This is worked out by the following procedure.

The percentage of the total reassociated DNA is plotted against the initial concentration of polynucleotide, C_0, multiplied by the time of reassociation, t (that is, C_0t), on a logarithmic scale. Plots for *Escherichia coli* and calf thymus DNA are shown in Figure 2. As can be seen, *E. coli* gives a single curve shaped like the mirror-image of an S (and called an S-shaped curve), whereas calf thymus DNA gives a curve which is a composite of two S-shaped ones (but with one end of each 'S' missing in Figure 2, as the curve does not extend far enough); this indicates that calf thymus DNA contains at least two components which reassociate at widely different rates. The size of each component is estimated from the position of the point of inflection, marked I, where the two S-shaped curves meet. This gives approximately 40 per cent which reassociates much faster than the main component comprising the remaining 60 per cent.

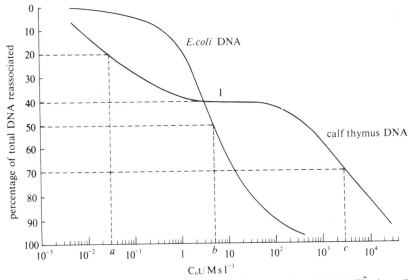

FIGURE 2   Reassociation curves for calf-thymus DNA and *E. coli* DNA. The letter I marks the point of inflection (at 40 per cent reassociation) in the calf-thymus curve of total DNA. The letters a, b and c on the horizontal axis mark the $C_0t_{(50)}$ (see text) of *E. coli* DNA, and the two main fractions of calf-thymus DNA.

In order to compare different curves, the value of C_0t at which 50 per cent of the polynucleotide strands are reassociated, or $C_0t_{(50)}$, is noted.

☐ From Figure 2, what are the values of C_0t corresponding to 50 per cent reassociation in (a) *E. coli* DNA, and (b) the most rapidly reassociating and the more slowly reassociating components of calf thymus DNA?

■ (a) $5\,M\,s\,l^{-1}$, (b) 0.03 and $3000\,M\,s\,l^{-1}$, respectively. (These are the points Marked b, a and c, respectively, in Figure 2. M stands for mole here.)

The $C_0t_{(50)}$ values are inversely proportional to the effective concentration of complementary DNA strands present, that is, the more copies of complementary DNA per unit volume, the faster the reassociation. In Figure 2, the fast reassociating component in calf thymus DNA has a $C_0t_{(50)}$ value 10^{-5} times that of the main fraction (0.03 compared with 3000). If we assume that the main fraction consists of single-copy sequences, then the fast component must contain sequences repeated 10^5 times. Other studies have shown that there is a third class of DNA which reassociates even faster—which must indicate that there is even more repetition. DNA of all three classes appears to be present in all organisms above the prokaryote grade; different organisms contain different proportions of these three classes of DNA (summarized in Table 1), which are distinguished by the extent of repetition.

repetitive DNA

TABLE 1   Three classes of DNA sequences found in eukaryotes

Class of DNA	Percentage of genome	Number of copies per haploid genome	Examples
unique (no repetition)	10–80	1 or 2	structural genes for haemoglobin, ovalbumin, and silk protein
middle repetitive	10–40	10^1–10^5	genes for rRNA, tRNA, and histones
highly repetitive	0–50	$> 10^5$	DNA sequences of 5–300 nucleotides

The production of repetitive DNA sequences is in all probability due to 'accidents' or events not directly related to their ultimate function. They could have arisen either from mechanisms generating tandem repeats, such as unequal chromatid exchanges which sometimes occur in mitosis (see Figure 3a) or unequal crossing over in meiosis (Figure 3b).

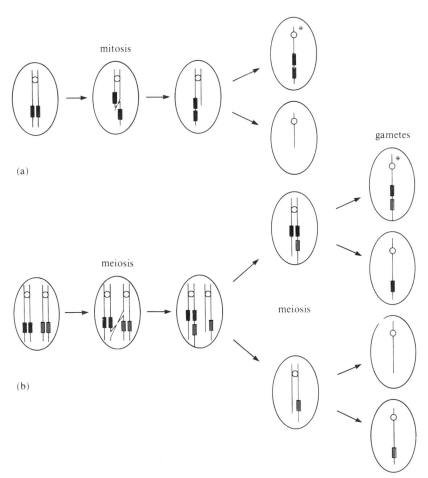

FIGURE 3   (a)   Unequal exchange of segments between sister-chromatids (chromatids formed by duplication of a chromosome) during *mitosis*, giving rise to a chromosome with duplicated segments as one of its products.

(b)   Unequal exchange of segments between chromatids on different homologous chromosomes during *meiosis*, giving rise to a chromosome with duplicated segments as one of the products. Cells with duplicated segments are marked with an asterisk.

11

Wholesale duplication of entire genomes, or polyploidy, is another mechanism for increasing the DNA content of cells. It has been suggested that a series of polyploidizations as well as saltatory tandem duplications occurred early in the phylogeny of the vertebrates, so that the fish and amphibian ancestors of mammals had already attained the characteristic genome size required for the different grades. Subsequent big leaps in evolution were accomplished by differential use of the already existing repetitive DNA sequences. Polyploidy is a frequent occurrence among plants and lower vertebrates but it is impossible to tell whether it was involved in the evolution of the tetrapods. The main point being made here is that increase in DNA or genome size (however accomplished) could be responsible for the macroevolutionary origin of new types or grades of organisms.

4.2 Repeated sequences and regulatory genes

Britten and Davidson (1969) proposed a model for gene regulation in eukaryotes which uses the repetitive sequences in the genome as regulatory elements. A hierarchy of interacting genes is brought into action through specific inducing molecules produced by control or regulatory genes. Thus a characteristic combination of enzymes and other proteins is produced when a particular control gene is activated. Differentiation would involve the sequential activation of distinctive batteries of genes in different cells. Two versions of the model for gene regulation, both making use of redundancy, are shown in Figure 4.

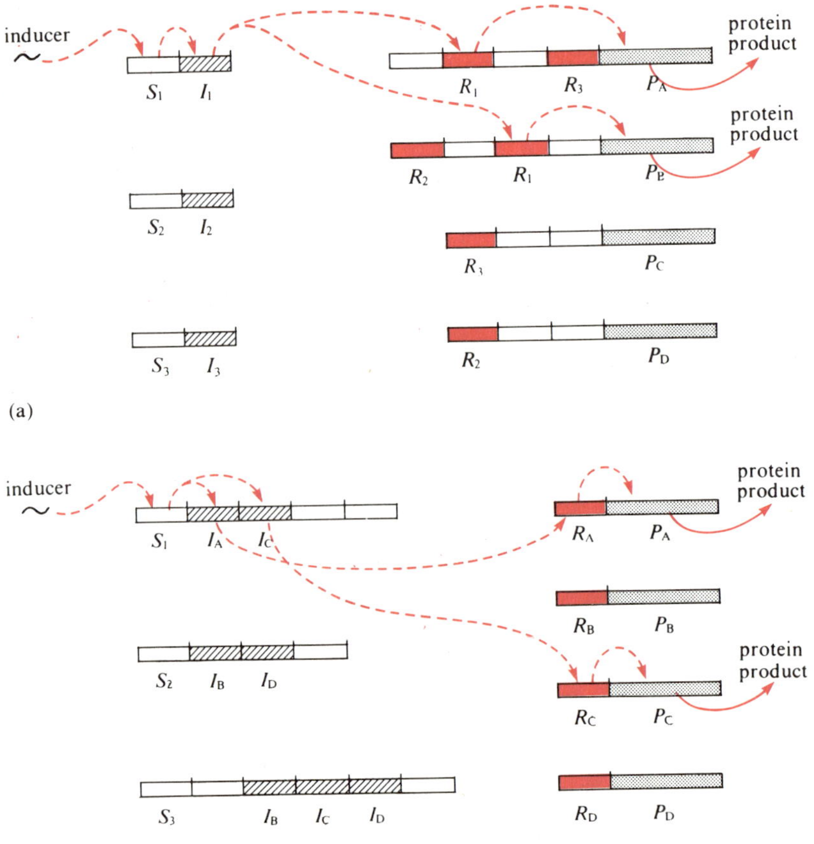

Britten–Davidson model (1969)

(a)

(b)

FIGURE 4   Two models of gene organization, each involving three chromosomes, which could give differential gene activation in different tissues. Dashed lines indicate activation. Activated sensor genes (S) cause sequential transcription of (i) linked integrator (I) genes, (ii) specific unlinked receptor (R) genes (shown here as having the same subscript letter as the I genes), and (iii) specific producer (P) gene(s) linked with the R gene(s) (shown here as having the same subscript as the R genes).

An inducer molecule, possibly a particular protein in the nucleus, binds to a specific *sensor gene*, S. This results in the activation and transcription of the *integrator gene(s)*, I, linked to the sensor gene, to give activator RNA molecule(s) which in turn bind to the appropriate *receptor gene(s)*, R. When the receptor gene is activated, transcription of the *producer gene(s)*, P, linked to the particular receptor gene takes place. The producer gene product is a messenger RNA or a messenger RNA precursor which is finally translated into a polypeptide.

sensor gene, S
integrator gene, I
receptor gene, R
producer gene, P

ITQ 4 (a) Identify the repeated sequences in (i) Figure 4(a) and (ii) Figure 4(b).

(b) Which of the producer genes would be activated when a specific inducer binds to S_3, in (i) Figure 4(a), and (ii) Figure 4(b)?

Britten and Davidson suggest that the likely candidates for the repeated receptor genes or integrator genes may be some of the middle repetitive DNA sequences described in Table 1, excluding those coding for RNA. Analysis of sequence organization in the DNA of *Xenopus* (African toad) and *Strongylocentrotus* (sea-urchins) has shown that the pattern of distribution of repetitive DNA is similar in the two organisms. Much of the genome consists of interspersed repetitive and unique sequences, which are about 300 and 1 000 nucleotide pairs in length respectively. This organizational pattern is consistent with the hypothesis that interspersed repetitive sequences somehow control the transcription of the unique sequences.

There is no doubt that the synthesis of proteins in different cell types is regulated in some way. Logically, therefore, the existence of mechanisms for controlling protein synthesis is a necessity. The *details* of the mechanisms involved, however, are still very hypothetical. Molecular biologists have recently turned up numerous surprises concerning the structure of the genes and the organization of the eukaryote genome. For example, structural genes coding for polypeptides are often not single continuous stretches of DNA. Each 'gene' may be split up into several parts separated by non-coding regions, i.e. stretches which do not code for any amino acid. Both the coding and the non-coding regions are transcribed as a unit, and then the transcribed RNA is further processed within the nucleus by enzyme systems which splice the coding regions together, snipping out the non-coding regions. The resulting messenger RNA is only then transported to the cytoplasm for translation. The transcription–translation process necessary for this is very complex. This means that the relationship between DNA and proteins becomes quite tenuous. There is certainly a lot of scope for regulatory mechanisms.

Davidson and Britten (1979) have modified their original model in the light of new knowledge concerning the eukaryote genome. The elements of sensor, integrator, and receptor and producer genes remain, but with the following alterations (see Figure 5).

Davidson–Britten model (1979)

(a) Every producer gene (*P*) with its receptor genes (*R*) is continuously transcribed as a unit (instead of being turned on specifically by the integrator gene products).

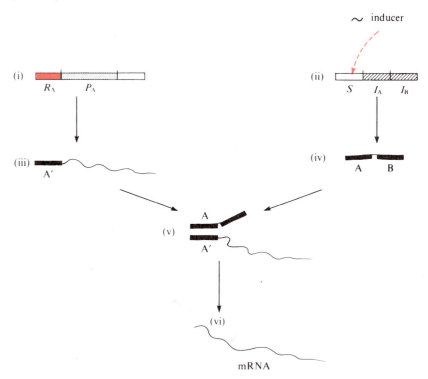

FIGURE 5 Model of the mechanism of differential gene activation. The receptor-producer gene unit (i) is continuously transcribed to produce a long stretch of RNA (iii). The sensor gene *S*, on being activated, causes the continuous integrator genes I_A and I_B (ii) to become transcribed to the RNA species (iv). Binding between complementary regions A′ and A of (iii) and (iv) gives the RNA–RNA complex (v), which facilitates processing of (iii) into messenger RNA (vi).

(b)   The RNA transcripts of integrator genes interact with the transcripts of certain producer–receptor units (which carry the appropriate receptors), forming RNA–RNA complexes. This in turn facilitates the processing of the producer–receptor transcripts into messenger RNAs which are transported to the cytoplasm for translation.

Circumstantial evidence does exist for general mechanisms of this kind. Indeed, one can predict that the next few years will see much unravelling of the genetic mechanisms for the control of protein synthesis. The real question is whether that will tell us anything concerning the relationship between genes and morphology.

The major difficulty that proponents of evolution face, is that of accounting for the origin and transmutation of organisms. The particular form of an organism arises during the process of development (or ontogeny), when the fertilized egg or spore cell undergoes a series of changes to become a complex multicellular adult. The essence of morphological evolution is the initiation of new phylogenetic lineages by alterations in the development or life histories of organisms. To discover the factors involved in this, we shall need to look more closely at the visible relationships between ontogeny and phylogeny, and develop a method for the analysis of form.

ontogeny

4.3   Objectives and SAQs for Sections 2 to 4

Now that you have completed these Sections you should be able to:

(a)   Define or recognize the meaning of the following: microevolution, macroevolution, quantum evolution, saltatory evolution, repetitive DNA, the Britten–Davidson models of gene regulation, neutralist theory, genome size, regulatory genes, gene duplication.

(b)   State the major difficulties involved in accounting for macroevolution in terms of microevolutionary processes and the models advanced to explain it.

(c)   Describe and interpret data relating to the model of gene regulation based on repetitive DNA.

To test your understanding of Sections 2 to 4, try the following SAQs.

SAQ 1 (*Objectives a and b*)   Why is it difficult to account for macroevolution in terms of microevolutionary processes?

SAQ 2 (*Objective c*)   (a) In Figure 4(a), which sensor gene(s) need to be induced, in order to obtain P_B protein? (b) In Figure 4(b), which producer genes are turned on when S_2 is induced?

SAQ 3 (*Objectives a and c*)   The crucial element in differential gene expression in Davidson and Britten's models as depicted in Figures 4 and 5, is the activation of different sensor genes in different tissues. Why can such a model never give an adequate picture of gene control during the whole of development?

5   Ontogeny and phylogeny

Since the very early days of embryology and comparative anatomy, the notion has existed that the individual, during its ontogeny, goes through a series of forms resembling its adult ancestors. In other words ontogeny recapitulates phylogeny or, as expressed popularly, each animal during development climbs up its family tree. An example of this phenomenon is the transient presence of gill pouches in the embryos of birds and mammals; these pouches appear to recapitulate those in adult fish—the ancestor of both birds and mammals (see Figure 6).

recapitulation

The most enthusiatic champion of this view was Haeckel, who elevated recapitulation to the status of a natural law in the 19th century. He claimed that the resemblance between the ontogenetic stages of an individual and the phylogenetic sequence was simply due to the fact that the individual descended from its sequence of ancestors. Von Baer, however, from his careful study of ontogenies, had previously noted that the young stages of different animals often resemble one

Haeckel's law

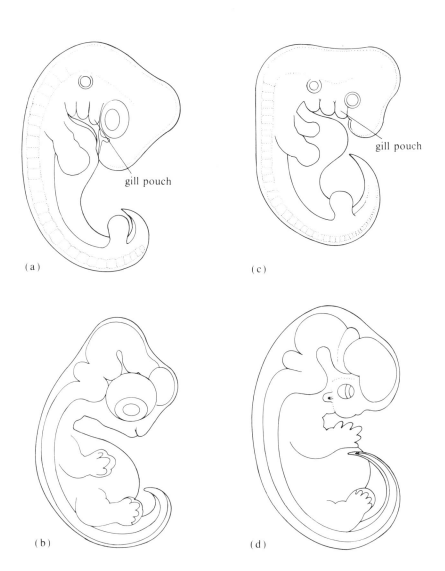

FIGURE 6 Embryos of (a) four day old chick, (b) eight day old chick, (c) four week old dog, (d) six week old dog.

another more than the adult stages resemble each other. He rejected recapitulation and put forward his famous laws of development as follows:

Von Baer's laws

1 In development from the egg, the general characters appear before the special characters.

2 From the more general characters, the less generalized and finally the specialized characters are developed.

3 During its development, an animal departs more and more from the forms of other contemporary animals.

4 The young stages in the development of an animal are not like the *adult* stages of other animals lower down the evolutionary scale, but are like the *young* stages of those animals.

Though there are a few exceptions (e.g. special characters related to the nutrition of embryos, such as the embryonic membranes, appear very early in development), these laws are accurate generalizations about the process of development. Therefore it is not common descent which has brought about the resemblance between the embryonic stages of different animals, but a general tendency towards progressive differentiation: all organisms whose adults have the same basic bodyplan have to pass through the same sequence of developmental stages.

The real difference between Haeckel and Von Baer is in their attitude towards evolution. Haeckel's law is explicitly evolutionary: resemblances are due to descent from a common ancestor (or heredity). Von Baer's laws of development, on the other hand, state that resemblances between ontogenies are due to internal constraints: there is no need to invoke a common ancestor. Can Von Baer's *observations* be given an evolutionary interpretation? Indeed, they can, as Darwin was the first to realize; the same conservative tendency of heredity could account for the resemblance between the embryonic stages of different animals. Even so, there may be laws of development that underlie some embryonic resemblances, quite independently of heredity.

5.1 Evolutionary synthesis

Earlier this century, Walter Garstang did much to translate Von Baer's laws into evolutionary terms. He pointed out that Haeckel had overlooked the fact that there is another evolutionary sequence, the evolutionary succession of zygotes, running more or less parallel with the adult sequence though steadily diverging from it. In the course of evolution, he suggested that 'an elaboration of zygotic structure' took place involving both the cytoplasm and the nucleus, and this gave rise to changes in adult structure (see Figure 7).

The real phylogeny of the metazoa has never been a direct succession of adult forms, but a succession of ontogenies or life cycles. Thus, Haeckel's cause and effect had to be reversed. Ontogeny recapitulates phylogeny, not because the latter causes the former, but because ontogeny *creates* phylogeny. Evolution is the history of the changes in development of organisms. Before we examine these changes, we need to deal with the related problem of biological form.

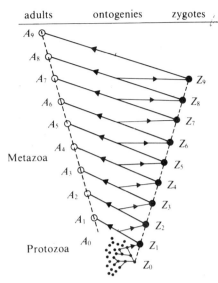

FIGURE 7 The relationship between ontogeny and phylogeny. A_0–A_9 demonstrates the evolutionary succession of adults; Z_0–Z_9 demonstrates the evolutionary succession of zygotes; Z_1–A_1, Z_2–A_2, etc., demonstrate the evolutionary succession of ontogenies.

6 Did archetypes exist?

Students of morphology (both evolutionist and anti-evolutionist) have long recognized the similarity of organs and 'design' in otherwise widely different species. Indeed, the poet Goethe, in his *Metamorphosis of Plants*, postulated that all existing land plants are derived from a common ancestor, the *Urpflanze* (arche-plant), and that all the organs of plants are homologous modifications of a single structure, the leaf. Such ideas gave rise to the concept of an archetype, which in an evolutionary context represents the ancestral type from which descendent species are derived (Unit 3).

☐ Does this mean that such an ancestral type necessarily existed?

■ No. From comparative studies of diverse species, morphologists derived a possible generalized bodyplan from which the more specialized features of present-day organisms could have been derived. The hypothetical organism having this basic bodyplan, they called the archetype (Unit 3).

The enormous diversity of organisms could all have evolved from just a small number of archetypes. Moreover as generalized features usually appear earlier in phylogeny than specialized ones, the fossil record gives some support to the concept of an archetype. However, few archetypes have been found in the fossil record so they should be regarded as hypothetical rather than real. Nevertheless, large groups of organisms *could* all have been derived from a simple archetype, as discussed for metazoan phyla in Unit 3.

The unity of types within each major grouping, such as a phylum, may be taken as evidence for (a) descent from a common ancestor, or (b) the existence of developmental laws such as those of Von Baer which necessarily constrain the paths of development. These two alternatives are not mutually exclusive.

☐ What conclusions may be drawn concerning the phylogeny of major groups if either (a) alone or (b) alone is true, or if *both* (a) and (b) are true?

■ If (a) alone is true, then each major group is likely to have evolved from a single common ancestor (monophyletically), because the origin of an ancestor would itself have been a rare and accidental event. If (b) alone is true, no common ancestor is necessary and different lines within the group may have had independent origins, that is, they could be polyphyletic. If both (a) and (b) are true, then the group could be either monophyletic or polyphyletic.

There appear to be many polyphyletic groups, especially among animals.

ITQ 5 Give two examples of polyphyletic groups of animals, from your reading of Units 5 and 6.

The existence of polyphyletic groups is itself the consequence of another widespread phenomenon—parallel changes in evolution (including parallel evolution, convergence and iterative evolution, see Unit 8, Section 2).

16

ITQ 6 Give two examples of parallelism in evolution, from your reading of Units 7–9.

If these striking parallelisms occurred solely by natural selection of random mutations, one has to postulate (i) that the selective pressures of the environments on the organisms were almost precisely the same, and (ii) that the mutations which took place in the organisms had nearly identical effects. Where long sequences of parallel evolution are involved, one has to postulate in addition that the changes in selective pressure and the different mutations both occurred in the same sequence for the different lineages. The chances of all these occurring together are remote. To some evolutionists, this speaks most eloquently for the existence of developmental laws and constraints which are independent of both the environment and of mutations in DNA. This was mentioned in relation to parallel evolution within the Mollusca (Unit 8, Section 2.1) where no 'adaptive' explanation of parallelism is feasible.

6.1 On growth and form

D'Arcy Thompson, in his classic book *On Growth and Form* published in 1917, was the first to make a rational approach to comparative morphology. He investigated the physical and mechanical *causes* of biological form and demonstrated how diverse shapes are interrelated by simple transformations.

His thesis is perhaps best illustrated by one of the examples he treats in some detail. Darwin regarded the honeycomb as the most wonderful product of instinctive action and declared that 'beyond this stage of perfection in architecture, natural selection could not lead: for the comb of the bee-hive, as far as we can see, is *absolutely perfect* in economising labour and wax'. Darwin's admiration was based on the fact that honeycombs have a great regularity of form. They consist of arrays of hexagonal cells (see Figure 8). However, D'Arcy Thompson showed that hexagonal arrays are almost invariably produced if uniform bodies with circular cross-sections are crowded together (see Figure 9). Thus, the beautiful regularity of a honeycomb, attributed to instinct perfected by natural selection, can largely be explained as the automatic outcome of the interplay of physical and mechanical forces.

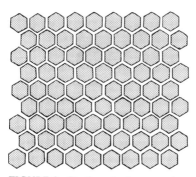

FIGURE 8 Portion of a honeycomb.

In Unit 1, it was explained that if a body increases in size and its shape remains approximately the same, the surface area increases as the square of the linear dimensions and the volume as the cube. This simple rule has important consequences for both the architectural design and physiology of living things, as described in Unit 1.

ITQ 7 Use this rule to explain why an elephant's legs are disproportionately broader (i.e. in cross-sectional area) than those of a mouse. Assume that the elephant is 50 times the length of the mouse.

Biological forms are therefore at least to some extent moulded by mechanical and physical forces and hence may be analysed in those terms.

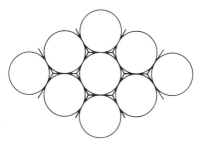

FIGURE 9 Production of hexagonal cells by pressing circular cells together.

6.2 The theory of transformation

We have seen that, in spite of the great diversity of living organisms, they can be divided into a small number of basic types. It is important for morphologists to find some means of comparing related forms as well as providing a precise description of each organism in isolation. The method which suggested itself to D'Arcy Thompson was that of *transformation*, a mathematical procedure in which the coordinates defining one shape are continuously deformed into those defining a different shape. This is like drawing a form on a rubber sheet, which can then be stretched or distorted to give the desired transformation into another form. Some examples of this procedure are given in Figure 10.

transformation

Figure 10 demonstrates that a regular mathematical relationship exists between certain forms and that, by a suitable combination of different rates of growth, one form can be transformed into another. It must be remembered, however, that these transformation diagrams cannot represent real phylogenetic transformation, for they represent *adults*; in phylogeny, transformations occur at the

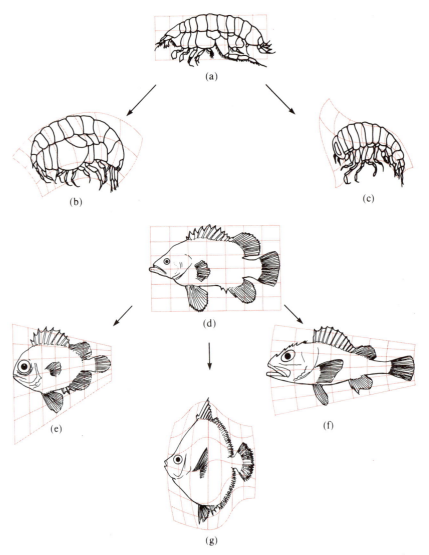

FIGURE 10 (a–c) Three different species of crustaceans, and (d–g) four different species of fish. In both (a) and (d) the animals are shown against symmetrical grids; (b) and (c), and (e)–(g), show that the forms of other species can be derived by transformation of these symmetrical grids.

embryonic stages. Nevertheless, these transformations demonstrate that the evolution of form can be analysed in terms of differential growth rates for different parts of the organisms. D'Arcy Thompson's two-dimensional transformation diagrams illustrate this concept superbly, but it would be difficult to analyse form in this way in practice. Therefore a much simpler method of analysis, based on *allometry*, is used (see Section 6.3).

allometry

6.3 Allometric growth

As D'Arcy Thompson himself pointed out, all but the simplest organisms reach their adult form or shape by differential growth in different directions.

ITQ 8 Why is this *necessarily* true?

This has stimulated the formation of empirical laws which appear to govern differential growth. (An empirical law is one which is based on observations, and which gives a useful description of a whole class of phenomena without going into any mechanisms involved.) The allometric equation

$$y = bx^k \qquad (1)$$

allometric equation

can be used to describe the relationship between the growth of a part of an organism to that of the whole organism. In this equation, y is the magnitude (usually weight or length) of the part, x is the magnitude of the whole body (or whole body less the part) measured in the same units as y, and b and k are constants. This equation is basically the same as the equation $E = kp^{0.66}$ given in Unit 6, Section 5.5.2, except that different symbols are used.

The constant b was once considered to be of no particular significance, for it simply denotes the value of y when $x = 1$, that is, the fraction of x which y occupies when x equals unity (remembering that $1^k = 1$). We shall say more about its real significance in Section 8.1. The coefficient k has an important meaning. It implies that, for the range over which the formula holds, the ratio of the specific growth rate of the organ to that of the body remains constant. By specific growth rate is meant the (instantaneous) rate of growth per unit weight. Equation 1 is usually written in the logarithmic form,

$$\log y = \log b + k \log x \qquad (2)$$

so that when x and y are measured in the same units and y is plotted against x on similar logarithmic scales, a straight line (see Figure 11) of slope k intercepting the y-axis at $y = b$ is obtained. (This may remind students of S101 of the graph of $y = mx + c$ in MAFS.) Note that the y-axis is chosen as the line $x = 1$, i.e. the line $\log x = 0$ (remember that $\log 1 = 0$). In the Figure, for both sets of axes, $k = \text{slope} = \text{rise/run} = 1.6$ (at the particular point where we have measured it on the Figure, $k = 32\,\text{mm}/20\,\text{mm} = 1.6$, but the same value would have been obtained whatever point had been chosen). The value for b, $0.0016\,\text{kg}$ ($= 1.6\,\text{g}$), can be read directly from the left-hand axis, or be derived from the $\log b$ value on the right-hand axis by using a calculator or log tables.

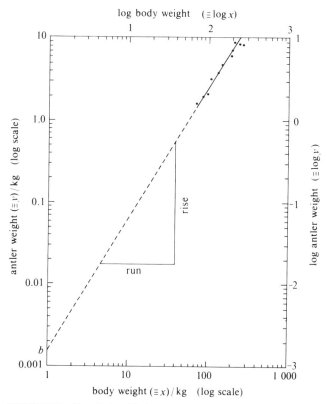

FIGURE 11 The relationship between antler weight and body weight in adult red deer, *Cervus edaphus*; both weights are plotted on a logarithmic scale, to give a straight line graph. The axes for $\log x$ versus $\log y$ are also given (top and right) for comparison.

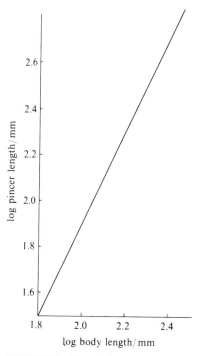

FIGURE 12 The relationship between the logarithm of pincer length and the logarithm of body length in the prawn, *Palaemon malcomsoni*, during development.

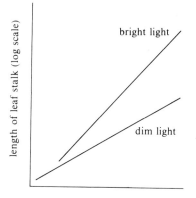

FIGURE 13 The logarithm of the length of the leaf stalk plotted against the logarithm of the diameter of the leaf, for nasturtium (*Tropaeolum*) leaves grown in bright and dim light.

Huxley suggested that equation 1, which implies a constant ratio between specific growth rates, might express a general law of differential growth. Indeed, it has been found to describe adequately a wide range of growth phenomena including:

1 variations in proportions correlated with variation in absolute size in adults of a single species (see Figure 11);

2 change in proportion during growth in animals and plants (see Figures 12 and 13);

3 change in adult proportion with increasing absolute size in related species of living animals, and in fossil species in the course of evolution.

4 changes in the proportions of various chemical constituents in growing organisms (see Figure 14).

Allometry means a change of proportion with increase in size, both within a single species during development and between adults of evolutionarily related groups.

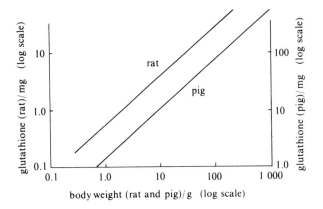

FIGURE 14 A double logarithmic plot of the increase in glutathione in rat and pig embryos during development.

The allometry is *simple* if the equation $y = bx^k$ holds for the entire range of values studied. *Isometry* is the term used when $k = 1$; this means that the geometric proportion remains unchanged with size (see Figure 15). When $k > 1$, there is *positive allometry*, in other words, the relative size of the organ increases as body size increases. *Negative allometry* holds when $k < 1$; in this case, the relative size of the organ decreases as total body size increases.

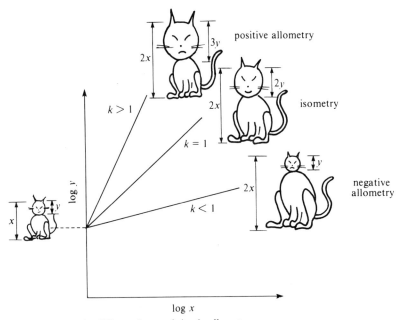

FIGURE 15 The different forms of simple allometry.

The allometric formula allows us to ignore the time factor involved in growth by relating the size of different parts of the body to each other, regardless of age. In other words form, during growth, is a function of absolute size rather than of absolute age, *in so far as size and age are independent of each other*. (We shall see in Section 8.1, however, that a knowledge of age, as well as size and shape, is crucial to the analysis of the mechanisms involved in changes of form in evolutionary lineages.) The general applicability of the allometric formula reflects the regular correlation of growth between different parts of an organism.

It is important to realize that there are three distinct uses of the allometric equation:

1   In *ontogenetic* allometry the correlation of change in shape with increase in size is measured during the *development* of organisms belonging to a single species.

2   In *intraspecific* allometry the correlation of variation in shape with *adult* size is measured for individuals belonging to a single species.

3   In *interspecific* or *phylogenetic* allometry the variation in shape with adult size is measured for *different species*.

A great deal of evolutionary change involves allometric alterations, that is, changes in shape or proportion. It is in the analysis of shape changes that the allometric formula is most useful.

One of the first applications of the allometric formula to phylogenetic data was that of Hersch in the 1930s, who analysed some extensive fossil data on the titanotheres. This remarkable group of ungulates of the Eocene and Oligocene was confined largely to North America. During their brief evolutionary history, these ungulates exhibited parallel size increases in several lineages, accompanied by striking allometric growth of bony, horn-like protuberances on the pre-orbital surface of the skull.

Titanothere horns began as a pair of lateral swellings in certain small Eocene forms that were ancestral to the three major Oligocene subfamilies. During the evolution of these subfamilies, the swellings tended to fuse at the base, lengthen, and diverge at the distal end to form a single V-shaped structure (see Figure 16).

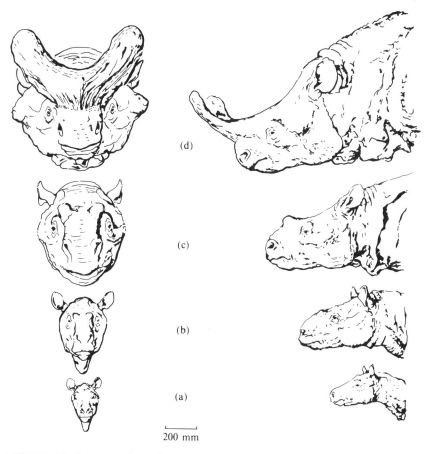

200 mm

FIGURE 16 Reconstructions of four fossil species representing successive stages in titanothere evolution (an approximate phylogenetic sequence). (a) *Eotitanops borealis*, Lower Eocene; (b) *Manteoceras manteoceras*, Middle Eocene; (c) *Protitanotherium emarginatum*, Upper Eocene; (d) *Brontotherium platyceras*, Lower Oligocene.

Growth of the horn relative to overall skull length exceeded the relative growth of almost every other recorded biological structure showing allometry. Comparing the length of the horn to the length of the skull, Hersch calculated a k value of about 9, a very high value.

Thus, striking parallel size increases took place, from small essentially hornless animals to large horned forms, in three independent lineages. This convinced Osborn that true orthogenesis (a non-adaptive trend arising from an intrinsic tendency within the organisms themselves) was involved. He suggested that the parallel trends could only be due to inherent tendencies in a common ancestor. Julian Huxley (1932) pointed out that natural selection for general body size increase could be sufficient to evoke the 'latent potentialities of the horn's growth-mechanisms' which would then proceed according to the allometric growth relationship. More recently, it has been suggested that sexual selection for head butting (used in competition for mates) probably contributed to the evolution of the titanothere horns.

orthogenesis

ITQ 9   Recall from Unit 8 an example of allometric size increase comparable to the titanothere horns.

Hersch also analysed a number of other allometric relationships in the titanotheres. He came to the following tentative conclusions:

1   Species within a genus all have the same values for *b* and *k*.

2   Genera are distinguished from one another by differences in *b* and *k*.

These changes in allometric relationships in evolution are of great interest in suggesting possible mechanisms involved in changes of form that are often regarded as macroevolutionary. One such mechanism is *heterochrony*, the temporal displacement of a character during ontogeny. We shall discuss this in Section 8.

heterochrony

Summary of Sections 5 and 6

1   Parallels exist between ontogeny and phylogeny.

2   Ontogeny provides the potential for phylogenetic change.

3   The diversity of form in the living world is the result of differences in ontogenies.

4   The resultant biological forms are subject to the operation of mechanical and physical forces as well as to architectural constraints on design, for effective functioning.

5   Regular transformations govern the change of form in phylogeny.

6   These regular transformations reflect an orderliness of growth patterns which may be analysed by allometry.

7   Allometric relationships characterize many phylogenetic transformations.

8   Allometric analyses suggest mechanisms whereby phylogenetic transformation may take place.

6.4   Objectives and SAQs for Sections 5 and 6

Now that you have completed these Sections you should be able to:

(a)   Contrast Haeckel's doctrine of recapitulation with Von Baer's laws of development, and state their evolutionary significance.

(b)   Define allometric growth and use the allometric equation (equation 1).

(c)   Distinguish between different allometric relationships.

To test your understanding of Sections 5 and 6, try the following SAQs.

SAQ 4 (*Objective a*)   What observations led Haeckel and Von Baer to formulate their respective laws?

SAQ 5 (*Objective a*)   Explain why Haeckel's law is explicitly evolutionary whereas Von Baer's laws are not.

SAQ 6 (*Objectives b and c*)   Examine Figure 17 carefully, then indicate whether the following statements are true or false.

(a)   The solid line represents the ontogenetic allometry of mammals in general.

(b)   Mammals in general have smaller brain weights than primates.

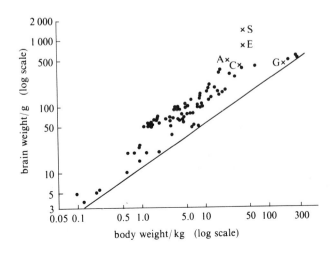

FIGURE 17   Relationship between brain weight and body weight among adult mammals in general (straight line) and primates (circles). C is the chimpanzee, G the gorilla, A the fossil hominid *Australopithecus erectus*, E represents the average for specimens of *Home erectus* (Java and Peking man) and S the average for modern humans (*Homo sapiens*).

(c) The gorilla has a smaller brain weight : body weight ratio than the chimpanzee.

(d) *Australopithecus* has a higher brain weight : body weight ratio than either of the *Homo* species.

(e) There has been a significant increase in brain size in hominid evolution.

7 The ontogenetic basis for phylogenetic change

More than 50 years ago, R. Goldschmidt suggested that the role of genes in morphogenesis was to control the rates of the processes involved. Mutations, by altering the velocity of one or several reactions relative to the rest, would shift the whole subsequent course of development. The result of such a systemic mutation would be a phenotypic change affecting a large number of tissues and organs that could form the basis of saltatory evolution (Section 3).

This physiological concept of gene function is an extremely important one, and has yet to be fully explored. A great deal of work has been done recently on the genetic control of protein synthesis (see Section 4.2), but almost nothing on the genetic control of timing in development. Goldschmidt described a number of examples which suggest that the timing of developmental processes is genetically controlled. We shall examine one of these in Section 7.1.

7.1 The genetics of pattern

The gypsy moth (*Lymantria dispar*) is distributed widely over Europe, China, Japan and part of the Atlantic coast of the United States. There is extensive formation of geographical races characterized by both physiological and morphological differences. One of these differences is the degree of pigmentation in the caterpillar, which varies from a very light to an almost black colour. Caterpillars may be grouped into ten classes on the basis of pigment intensity, the darkest in class I and the lightest in class X. Young caterpillars of different races show a characteristic frequency distribution among the classes (see Table 2).

TABLE 2   Percentage frequencies of colour classes I–X for eight races of gypsy moth

Race	I (darkest)	II	III	IV	V	VI	VII	VIII	IX	X (lightest)
	Frequency of colour class/per cent									
H								38.6	54.5	6.9
K							15	64.7	20.3	
F						1.5	60	38.5		
O						8.2	48.8	31.4	11.6	
G							11.8	57	31.2	
A					19	43	38			
S +		5.4	19.1	44	30.2	1.3				
MS	← 100 →									

Of great interest is the variation in the rate at which pigment becomes deposited during the development of the larva through its six moults. Three racial types are observed:

(a) *Light races*   These caterpillars remain light throughout larval life.

(b) *Intermediate races*   These begin as light to medium-light caterpillars, grow progressively darker with every moult, and finally are medium to dark.

(c) *Dark races*   Young caterpillars are medium-dark to dark, and remain dark throughout larval life.

Examples of the distribution of caterpillars among the different colour classes for some intermediate races are shown in Table 3.

☐ In Table 3, races H, G and A are listed in the order of a generally increasing rate of pigment deposition during development. What other factor is involved in determining the coloration at any stage?

TABLE 3 Examples of shifts in colour class frequency (in per cent) during development, for three intermediate races of caterpillar

Stage of development (no. of moults)	I (darkest)	II	III	IV	V	VI	VII	VIII	IX	X (lightest)
Race H										
3								38.6	54.5	6.9
4							11.8	55.3	31.7	1.2
5	29.3	24.4	9.7	12.2	9.8	12.2		2.4		
6	64	16	16	4						
Race G										
3							11.8	57	31.2	
4					1.4	35.2	33.8	28.2	1.4	
5	3.6	7.2	14.4	18	25	22.8	9			
6	←———100———→									
Race A										
3					19	43	38			
4					35.5	58.1	6.4			
5	14.3	42.9	4.7	9.5	28.6					
6	←———100———→									

■ The time of *onset* of pigment deposition.

The genetics of pigmentation is complex; the key observations are:

1 The colours of F_1 hybrids between a light race and a dark race tend to be intermediate between the parental colours. In the early stages (Figure 18a) they are nearly as light as the lighter parental race, but by progressive stages the caterpillars become darker (Figure 18b), though even after the sixth moult they tend to be lighter than the darker parental race.

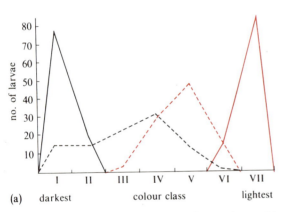
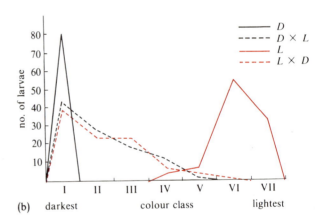

FIGURE 18 Distribution among the different classes of larval pigmentation of a dark race D and a light race L, and of the reciprocal F_1 generations of crosses between them. $D \times L$ represents the F_1 with a D female crossed with an L male, and $L \times D$ the F_1 with L female and D male. (a) Distribution after the second moult. (b) Distribution after the fourth moult. Note the shift in distribution towards the dark classes between the second and fourth moult. Note also that the larvae at first (in (a)) show a greater resemblance to the *maternal* parent.

2 F_2 generations obtained from F_1 hybrids initially exhibit an approximate ratio of 3 light to 1 dark (Figure 19a), but roughly two-thirds of the light caterpillars shift over into the dark classes in successive moults, thus reversing the ratio of light to dark larvae (Figure 19b).

3 The deposition of pigment during larval life proceeds at a more or less constant rate which varies between the different races (see Figure 20). A prolongation of a moulting stage (e.g. by starvation) consequently produces darker larvae than normal for the race at that particular stage of development.

4 In different races, pigment deposition begins at different times. The character controlling the timing of pigment deposition is inherited independently of the character controlling the presence or absence of pigmentation.

These observations show that pigmentation is a character which is controlled by at least two systems of genes. The first system, probably a single locus, determines

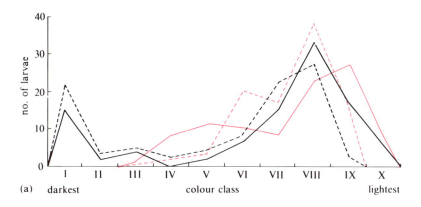

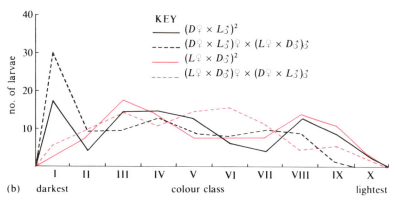

FIGURE 19 Distribution among the different classes of larval pigmentation in the F_2 generations of crosses between a dark race D and a light race L. Reciprocal F_1 crosses were made between the races; that is, (using ♀ to represent female and ♂ male) $D♀ × L♂$ and $L♀ × D♂$. From these, the four possible F_2 generations are: selfing of each reciprocal F_1 generation—$(D♀ × L♂)^2$ and $(L♀ × D♂)^2$; and intercrosses between the F_1 generations—$(D♀ × L♂)♀ × (L♀ × D♂)♂$ and $(L♀ × D♂)♀ × (D♀ × L♂)♂$. (a) Distribution after the second moult. (b) Distribution after the fourth moult. Note the shift in distribution towards the dark classes between the second and fourth moult. Note also that the early larvae (in (a)) show a greater resemblence to the *maternal* parent, indicating a cytoplasmic influence on pigmentation.

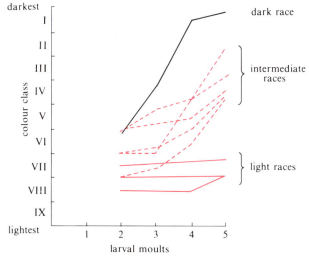

FIGURE 20 Changes in pigmentation during larval life, for dark, intermediate and light races.

the presence or absence of pigment; the second, a polygenic system (Unit 10), determines the rate and the time of onset of pigment deposition. An additional source of influence is cytoplasmic, as is evident from the data presented in Figures 18 and 19; the young hybrid larvae show a greater resemblence to the maternal parent—the one contributing the greater amount of cytoplasm.

cytoplasmic inheritance

These results clearly demonstrate the existence of separate genetic mechanisms that control the timing and rate of developmental processes such as pigment deposition. Alterations in these mechanisms could therefore lead to change of the type called heterochronic change, which is discussed in the next Section.

8 The principle of heterochrony

Study comment This Section is rather long and may present some difficulties for those unaccustomed to algebra. You are required to understand the principles of how to measure geometric similarity, but not to reproduce the derivation. If you have forgotten about logarithms and indices, consult the relevant part of the Foundation Course (see Table A1). A calculator and/or log tables will be needed.

In general, heterochrony results from the acceleration or retardation of *somatic development* versus reproductive development. (Somatic development includes all structures and functions not directly involved in reproduction.) The morphological results of heterochrony are either *recapitulation* or *paedomorphosis*. Recapitulation refers to the repetition of the adult characters of an ancestor in the embryonic or larval stages of its descendents. Paedomorphosis refers to the retention of juvenile features in sexually mature individuals. The points are summarized in Table 4.

somatic development

recapitulation

paedomorphosis

TABLE 4   The categories of heterochrony

Timing of development		Morphological result
somatic	reproductive	
accelerated	—	recapitulation
—	accelerated	paedomorphosis
retarded	—	paedomorphosis
—	retarded	recapitulation

There are three fundamental processes in development which are crucial to the study of the relationship between ontogeny and phylogeny, namely: increase in *size*; change in *shape* (including cellular differentiation); and reproductive *maturation*. Growth and morphogenesis are two aspects of somatic development. All three processes begin at different times and proceed at different rates relative to one another. Dissociation between the timing of any two of the three processes would result in heterochrony.

We have seen that allometry is the study of the change in the proportions or shape of an organism with increase in overall size. Using the simple allometric equation $y = bx^k$, we can investigate if heterochrony has occurred.

The procedure is as follows. Take the standard logarithmic plot of body size versus part size during ontogeny in an ancestral species (see Figure 21, line A). Designate a point S at a well defined developmental stage, usually the adult, as the standard of comparison for 'shape'. 'Shape' in this case is taken as the size of the part relative to body size. Draw a line of slope $k = 1$ through the point S (the red line, I, in Figure 21). This intercepts the y-axis at a point b'. From equation 2, the equation of this line is:

$$\log y = \log b' + \log x$$

or

$$y = b'x^1$$

Thus

$$\frac{y}{x} = b' \quad \text{(a constant)} \tag{3}$$

Since $k = 1$, the line I is the line of isometry (see Section 6.3); the shape defined by y/x is constant all along the line. If this same shape occurs at a smaller body size in the descendent species, we have acceleration of shape in relation to size. Conversely, if this shape occurs at a larger body size, we have retardation.

acceleration

retardation

☐ In Figure 21, two descendent ontogenies B and C are compared with the ancestral ontogeny A. Which of the following statements are correct?

  (i)   B is retarded with respect to A.

  (ii) C is accelerated with respect to A.

  (iii) B is accelerated with respect to A.

  (iv) C is retarded with respect to A.

■ (i) and (ii) are correct. The line representing the ontogeny of B crosses the line of isometry (I) at a larger body size than A, whereas that representing the ontogeny of C crosses I at a smaller body size.

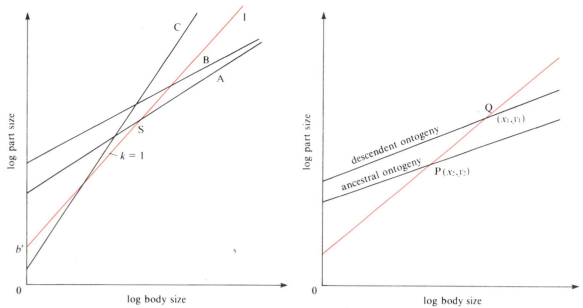

FIGURE 21 A demonstration of heterochrony by comparing the ontogenies of descendants with that of the ancestor.

FIGURE 22 A measure of heterochrony where ancestral and descendent ontogenies are parallel.

Often, ancestral and descendent ontogenies exhibit the same slope on such a graph (i.e. have the same k value) but intercept the vertical axis at a different point (i.e. have a different b value). In this case, a measure of the extent of heterochrony may be made directly without specifying a standard point for comparison. In Figure 22 the ontogenies of ancestor and descendant are plotted. The line of isometry through any point, say P, on the line representing the ontogeny of the ancestor intersects the line of descendent ontogeny at Q. P and Q are points representing the same shape, i.e. from equation 3

$$\frac{y_1}{x_1} = \frac{y_2}{x_2}$$

that is,
$$\frac{y_1}{y_2} = \frac{x_1}{x_2} \tag{4}$$

where (x_1, y_1) and (x_2, y_2) are the values of x (body size) and y (part size) at the points Q and P respectively. Now, according to the allometric equation (equation 1)

$$y_1 = b_1 x_1{}^k$$

$$y_2 = b_2 x_2{}^k$$

Therefore
$$\frac{y_1}{y_2} = \frac{b_1 x_1{}^k}{b_2 x_2{}^k} \tag{5}$$

Substituting for y_1/y_2 from equation 4,

$$\frac{x_1}{x_2} = \frac{b_1 x_1{}^k}{b_2 x_2{}^k}$$

Rearranging,

$$\frac{x_1 x_2{}^k}{x_2 x_1{}^k} = \frac{b_1}{b_2}$$

$$\frac{x_1 x_1{}^{-k}}{x_2 x_2{}^{-k}} = \frac{b_1}{b_2}$$

$$\left(\frac{x_1}{x_2}\right)^{1-k} = \frac{b_1}{b_2}$$

$$\frac{x_1}{x_2} = \left(\frac{b_1}{b_2}\right)^{1/1-k}$$

x_1/x_2 is the desired quantity—the ratio of the two body sizes at which shape is the same. This quantity we shall call the *coefficient of geometric similarity*, *s*. Thus,

$$s = \left(\frac{b_1}{b_2}\right)^{1/1-k}$$

The coefficient *s* (or x_1/x_2) simply defines the relative sizes at which the ancestor and the descendant have the same shape. For example, if $s = 5$ then the descendant has the same shape as the ancestor when it is five times the size. Its shape is thus retarded with respect to size.

ITQ 10   The allometric equations for ancestor (a) and descendent (d) ontogenies were as follows:

$$y_a = 0.237\, x_a^{1.766} \qquad y_d = 0.210\, x_d^{1.766}$$

(a) Calculate the coefficient of geometric similarity, *s*, for these organisms.

(b) What is the implication of the value for *s* found in (a)?

8.1   Standarization in the measurement of heterochrony

The use of allometry in the analysis of heterochrony suffers from the major defect that the dimension of time is ignored. By plotting different parts of the organism against each other, we get information about the dissociation between the general rate of growth and the *differential* rates of growth for the parts concerned. In other words we demonstrate a dissociation between *size* (due to general rate of growth), and *shape* (due to differential rates of growth of the parts) in descendent ontogenies as compared with that of the ancestor. In order to work out the *result* of this process of dissociation, however, we need to compare shape, size and *developmental age*. An example will make this clear.

Evolution of Gryphaea

The Jurassic oyster, *Gryphaea*, is to palaeontology what *Drosophila* is to genetics. It was long considered to be a classic example of an evolutionary trend in which the coiling of the lower valve of the shell increased to such an extent that the shells could no longer open and the lineage became extinct. This supposed trend was thought to have occurred in a number of lineages, and to have been accompanied in each case by an increase in size. Many biologists, inclined to see the working of natural selection everywhere, pointed out that a larger, more tightly coiled shell represented an adaptation for life in muddy bottoms, because the opening part of the shell, through which the animal feeds, would be raised clear of the mud. It was argued that since shells increase in size and coiling during ontogeny, an individual dying in old age because its shell would not open might nevertheless have been better adapted earlier in life than the less coiled individuals. There would thus be a balance between selection in favour of a high degree of coiling when young and against coiling in old age. As most individuals die before old age anyway, it was reasoned that this would result in changes advantageous to juveniles at the expense of the aged.

The data on which the whole story was based have been re-examined recently in the light of new fossil evidence. It was found that, although the increase in size is well documented in the fossil record, the same cannot be said for the increase in coiling. (This shows how easy it is to invent appropriate 'adaptive' stories to explain any character or trend.) Instead, *Gryphaea* began its post-larval life as a flat attached juvenile, coiling starting only when the juveniles became detached.

☐ Assuming that an allometric relationship exists between the degree of coiling and size, and that the relationship remains unchanged in descendants, what effect would the larger size of the descendants have on the degree of coiling?

■ There would be an increase in the degree of coiling with size. An estimate of the degree of coiling can be made by measuring the ratio of the perimeter (*p*) of the coiled valve to the length (*r*) of the flat valve (see Figure 23).

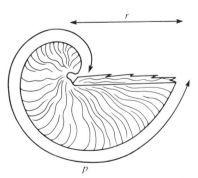

FIGURE 23   *Gryphaea incurva* from Lower Jurassic rocks of Britain, showing the measure of coiling—the ratio of the perimeter of the coiled valve (*p*) to the length of the flat valve (*r*).

This ratio was used to analyse the fossil data of *Gryphaea* and, much to the surprise of palaeontologists, showed that the descendants are *less* coiled than are ancestors of the same size. However, as coiling increases in ontogeny and size increases during evolution, these results could be an artefact of comparison between tightly coiled adult ancestors and more loosely coiled juvenile descendants. (Remember that valid comparisons can only be made between organisms of the same developmental age.) From the data, two equations relating *p* to *r* for the 'ancestor' and 'descendant' in fossil samples from two successive strata were constructed from the allometric plot of log *p* versus log *r*:

$$p_a = 0.237r_a^{1.766} \qquad \text{(for ancestor)}$$

$$p_d = 0.210r_d^{1.766} \qquad \text{(for descendant)}$$

ITQ 11 (a) Put the above equations in logarithmic form. (b) In these equations, the index has the same value, 1.766, for both ancestor and descendant. What does this tell you about the slopes of the two ontogenetic curves?

As the slopes of the ontogenetic curves are equal in the ancestor and descendant, it was possible to measure heterochrony by calculating the coefficient of geometric similarity *s*. We have already done the calculation towards the end of Section 8, and obtained a value of 1.17. As explained in the answer to ITQ 10, this means that descendants do not attain the same shape as the ancestors until they are 1.17 times as large; in other words, shape (coiling in this case) is retarded with respect to size. However, as the descendent *Gryphaea* are about 1.2 times as large as the ancestors, geometric shape is maintained—there is neither increase nor decrease in coiling during phylogeny. This example illustrates the pitfalls of standardization by size alone: in order to decide whether paedomorphosis or recapitulation has occurred (see Section 8), an appropriate biological standard such as age or developmental stage must be used.

The different intercepts on the vertical axis (corresponding to the logarithms of the constant *b* in the allometric equation) deserves some comment here. The *smaller* intercept in the descendant suggests that for the same initial body size the organ size is diminished, indicating a possible delay in the *onset* of organ growth. This can only be ascertained, however, from complete developmental data.

8.2 A clock model of heterochrony

How can we represent the results of heterchrony so that the processes involved are made clear? Gould (1977) devised a 'clock model' in which both results and processes are made explicit. The television programme 'Time for change' explains Gould's model and gives examples of its use. In this model, a semicircular clock is set up with two hands moving from left to right over two scales representing *shape* and *size* respectively. A third scale for *age* appears along the bottom of the clock (see Figure 24).

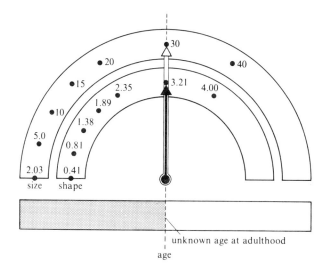

FIGURE 24 Scales of the clock model calibrated for *Gryphaea*. Adulthood is chosen as the standardization stage of the midline values, and grey shading indicates the juvenile period.

The ancestor and the descendant are compared at a standardized developmental stage—say the attainment of adulthood or sexual maturity. The age scale is calibrated in such a way as to place the ancestor's standardized adult stage at the midpoint of the scale. The initial age is recorded at the origin (to the left), and intervening ages are marked between the origin and the midpoint and then extrapolated beyond the midpoint. Juvenile stages (to the left) are shaded.

Next, from measurements of ancestral fossils, the scales of size and shape are matched to the age scale. This means that the two hands of the clock will move together when the ancestor is considered. Thus at the standardized developmental stage, the two hands of the clock and the grey shading at the base (indicating age attained) will all lie at the midpoint when the ancestor is considered (see Figure 24).

We are now ready to compare the descendent ontogeny with the clock calibrated against the ancestor, to see whether size and shape have become dissociated in the descendant. We want to know where the markers of shape, size and age in the descendent ontogeny lie on the ancestral scales, when the descendant has reached the developmental stage chosen for comparison. To see this, we divide the scales into two parts separated by the midline (see Figure 25).

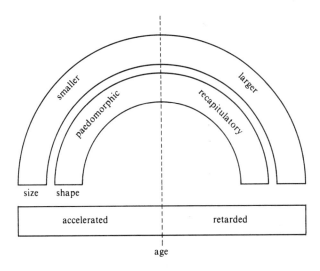

FIGURE 25 The domains of heterochrony—accelerated and retarded on the age scale, paedomorphic and recapitulatory on the shape scale, and smaller or larger on the size scale.

If heterochrony has occurred, changes will become apparent in any or all of the three domains of shape, size and age. For example, the hand for descendent shape may not lie on the midline as it did for the ancestor at the same developmental stage. If it lies between the origin and the midline, paedomorphosis has occurred because the adult descendant displays the shape of the youthful stages of the ancestor. If it lies beyond the midline, we have a case of recapitulation because the descendant has gone beyond the ancestral shape for the same developmental stage.

Similarly, the descendant may be either smaller or larger (to the left or right of the midline on the size clock, respectively), than the ancestor. We have accordingly a phyletic decrease or increase in size.

Again, the descendant may reach the chosen developmental stage at an earlier or a later time than the ancestor, that is, to the left or right of the midline in the scale for age. We have, accordingly, an accelerated or retarded development or maturation.

The results of heterochrony in relation to the processes can thus be classified according to the positions of the descendent ontogeny on all three scales (see Figure 26). Four basic types of heterochrony appear on the clock model, under two main headings.

(a) *Paedomorphosis*

(i) *Paedomorphosis by progenesis* (acceleration of maturation with respect to somatic development)—Figure 26(a). The descendent ontogeny is simply shortened or truncated by the early attainment of sexual maturity. The descendant is both smaller and paedomorphic—in other words, it is sexually mature but has the juvenile form of the ancestor. Size and shape remain correlated.

progenesis

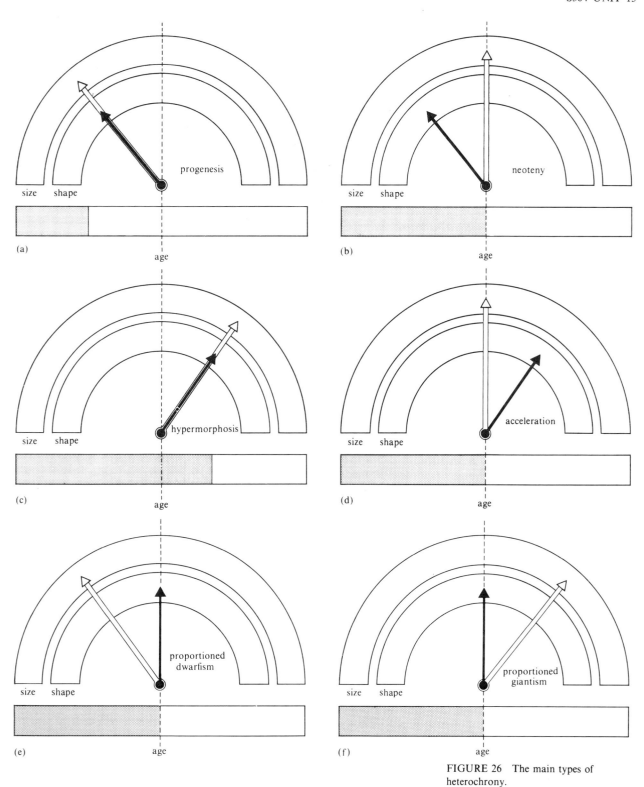

FIGURE 26   The main types of heterochrony.

(ii)   *Paedomorphosis by neoteny* (retardation of shape or somatic development with respect to developmental stage)—Figure 26(b). Here shape is retarded but size and age of maturation remain unchanged from the ancestral condition. The adult therefore *retains* juvenile characteristics.

neoteny

(b)   *Recapitulation*

(iii)   *Recapitulation by hypermorphosis* (retardation of maturation with respect to somatic development)—Figure 26(c). The correlation of size and shape remain unchanged from the ancestral condition, so the adult is larger and more highly developed.

hypermorphosis

(iv)   *Recapitulation by acceleration* (acceleration of shape with respect to developmental stage)—Figure 26(d). This is the classic case considered to be nearly universal by Haeckel. The age of sexual maturity remains the same but the somatic

acceleration

31

development proceeds faster in the descendant so that it surpasses that of the ancestor. The shape of the adult descendant is thus more highly developed than that of the ancestor.

Although a pure alteration in size does not give the main results of heterochrony (paedomorphosis or recapitulation), it is part of the general phenomenon of temporal dissociation. A retardation in the rate of size increase (Figure 26e) produces a dwarf, while an acceleration (Figure 26f) produces a giant; both these forms are geometrically similar to the ancestor.

proportioned dwarfism or gigantism

The clock model assumes that the data for age, developmental stage, size and shape are complete or nearly so. In practice, fossil data can rarely be classified by age. In the *Gryphaea* data discussed in Section 8.1, age is not known. Although we know that descendent *Gryphaea* were bigger than their ancestors but maintained the same shape at adulthood, we do not know how the increase in size was achieved. Either it was by an acceleration of growth without a concomitant acceleration in the rate of change in shape, or the size increase may have been achieved by a delay in maturation with a compensatory retardation in the rate of shape change. Thus, with the data for age missing, we have a restricted model which considers only size and shape (see Figure 27). From the data in the Figure, we judge that paedomorphosis has occurred but we cannot specify the processes involved.

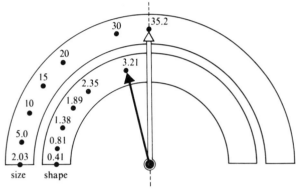

FIGURE 27 Restricted clock model for *Gryphaea*, considering only size and shape. Shape is retarded with respect to size, and we are forced to make a possibly false judgement of paedomorphosis by neoteny.

8.3 Heterochrony and evolution

Heterochrony is thought to underlie many evolutionary changes that may be classified as macroevolutionary. As you will recall, the two main results of heterochrony are paedomorphosis and recapitulation.

ITQ 12   How many kinds of paedomorphosis are there?

An intimate relationship between neoteny and macroevolution has been suggested by many workers. It has been linked to the origin of many higher taxa, e.g. insects and vertebrates. Indeed, the origins of *all* phyla, from protists to higher plants, have been linked to neoteny. The arguments involved are as follows:

1   As larval stages are almost never preserved as fossils, the occurrence of neoteny in evolution would give the impression in the fossil record that a large morphological change had taken place, whereas only a minimum of genetic change had been required (for example, a change in regulatory genes, see Section 3.2). Recall that there is some evidence for the occurrence of neoteny among the ammonite lineages in your Home Experiment.

2   If novel features were indeed introduced into the juvenile stages while adults remained in their ancestral condition, then neoteny would produce even larger morphological discontinuities in the fossil record.

All theories concerning the evolutionary significance of paedomorphosis have been based on morphological studies alone. Even accepting that reconstructions of morphological transformations are plausible, they serve only to identify the phenomena without supplying a casual explanation for their occurrence. The question which needs to be answered is this: Of what *immediate* significance to the ancestors themselves were the processes of heterochrony which led to progenesis or neoteny? Gould suggests that the clue may lie in the adaptive effects of different life-history strategies.

The factors which control the evolution of life histories had been studied for a long time before they were formalized in the theory of *r*- and *K*-selection of MacArthur (see Unit 11).

ITQ 13   What are *r*-selection and *K*-selection?

The theory suggests that alterations in life-history strategies would have an effect on *r* or *K*. In order to increase *r*, an organism can either devote more resources to reproduction (increasing the number of seeds or eggs), or *shorten its maturation time*. Conversely, an organism can reduce *r* and improve its competitive ability (possibly leading to increased *K*) either by minimizing the allocation of resources to reproduction or by a *retardation of maturation*. Clearly, heterochrony could drastically alter the population parameters of *r* and *K*. The *r*-strategy could be achieved by progenesis, and the *K*-strategy by neoteny. Thus, ecological factors favouring *r*-selection would be expected to be correlated with progenesis and factors favouring *K*-selection would be correlated with neoteny. It is premature to judge the validity of this because appropriate field data do not yet exist, but the idea is an interesting one.

ITQ 14   What sort of environments might favour progenesis and neoteny respectively?

8.4   Objectives and SAQs for Sections 7 and 8

Now that you have completed these Sections, you should be able to:

(a)   Define and recognize correct definitions of the terms heterochrony, paedomorphosis, recapitulation, progenesis, neoteny, hypermorphosis, proportioned dwarfism or gigantism, acceleration or retardation of shape.

(b)   Recognize data suggesting the existence of genetic controls in the timing and the rates of developmental processes.

(c)   Draw inferences concerning processes involved in heterochrony from allometric data.

(d)   Draw and interpret clock models of heterochrony and state the morphological consequences of different heterochronic processes.

To test your understanding of Sections 7 and 8, try the following SAQs.

SAQ 7 (*Objectives a, c and d*)   The ontogenetic allometry of body circumference (*c*, in mm) measured at the widest level versus length (*l*, in mm) measured from the tip of the snout to the base of the caudal fin, was compared in two species of fish, A and B, belonging to the same genus. The plots of log *c* against log *l* are shown in Figure 28.

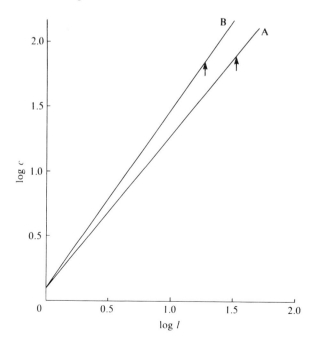

FIGURE 28   Plot of log *c* versus log *l*, for species A and B, where *c* and *l*, the circumference and length of the fish respectively, are in millimetres. Arrows indicate the point at which sexual maturity is reached in each species.

(a) From Figure 28, determine the values of the constants b and k and write down the logarithmic form of the allometric equation for circumference and length of body, for species A and B respectively.

(b) What single parameter has changed in the evolution of species B, assuming that A is more closely related to the common ancestor of the two species?

(c) What is the size (l) and shape (defined by c/l) of each species at sexual maturity?

(d) Is the shape of B retarded or accelerated with respect to A?

(e) Assuming that both species require 200 days to reach sexual maturity, draw a clock diagram representing the heterochronic changes in B compared with A, and interpret the data.

SAQ 8 (*Objective b*) If it could be shown by suitable breeding tests that the allometric growth of body circumference with increase in body length in the two species A and B in SAQ 7 is genetically controlled, what kind of genes would be involved, and what changes would have occurred in the evolution of species B?

SAQ 9 (*Objectives a, c and d*) Figure 29 gives the ontogenetic allometric curves for three species of primates. From these data, what is the most probable explanation for the observed increase in brain size during human evolution?

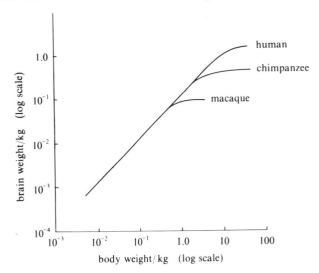

FIGURE 29 Ontogenetic allometry of brain weight versus body weight for three primate species.

9 Environment and change

In the last Section, we examined how changes in the rates of different processes of development (heterochrony) could generate large changes in morphology. It was suggested in Section 8.3 that certain environmental conditions may be correlated with different heterochronic changes, but no attempt was made to discuss how these and other environmental factors could influence development. This is an area of evolutionary studies which has been relatively neglected. It deals with the general relationship between environment and heredity in the evolution of adaptations.

We have seen that the neo-Darwinian theory explains adaptations in terms of the natural selection of random genetic mutations. Random mutations lead to the production of different phenotypes: those which happen to be adaptive in the environment, in the sense of enabling the individuals to survive *and* reproduce more successfully, will soon supplant others. In this way, one mutant allele replaces other alleles in the population. The environment is often seen as nothing more than the agent of selection, selecting out the 'good' genes from the 'bad', both of which were produced by chance. The organisms are in effect regarded as the 'programmed' products of randomly assembled genotypes.

Most biologists today regard such a view of evolution as inadequate. Phenotypes are generated by a continual interaction between the organism and its environment. At the most basic level, the zygote needs a suitable environment in which to develop. The development of individuals exhibits a kind of stability within a certain range of environmental conditions, so that essentially the same phenotype (or a restricted range of phenotypes called the *norm*) results. This phenomenon is referred to as *developmental homeostasis*. The range of environments within which developmental homeostasis can operate varies from one organism to another. Outside this range, small or large changes from the norm are invariably produced. This is referred to as *plasticity*. If the end result is a phenotype adapted to a particular environment, than the same phenomenon is called *developmental adaptability*. Thus, it comes as no surprise that the same genotype often gives rise to different phenotypes, depending on the environment.

developmental homeostasis

phenotypic plasticity
developmental adaptability

> **ITQ 15** In the television programme 'Species and evolution' you were shown an organism whose phenotype changed dramatically in different environments. Can you recall this example?

It is perhaps less often realized that the converse is also true: essentially the same phenotype can be generated by different genotypes. This time, it is an example of developmental homeostasis regulating against environmental fluctuations. The developing system as a whole regulates against *genetic* disturbances resulting from differences in alleles at various gene loci in such a way that the end result remains roughly the same.

☐ In the theory of evolution by natural selection of alternative alleles, what difficulties are created by the complex relationship between genotype and phenotype in different environments?

■ There are two main difficulties:
(i) The complex one-to-many and many-to-one correspondences between genotype and phenotype make natural selection of alternative alleles difficult to accomplish except under very special circumstances. At best we can envisage the natural selection of developmental systems as a whole (or whole genotypes, see Unit 10, Section 5.1, and Unit 12).

(ii) The sensitivity of the developmental process to environmental changes means that, in practice, it is difficult to say whether a novel phenotype is the result of a developmental reaction to new environmental conditions or of natural selection of a random mutation.

9.1 Developmental plasticity and adaptability

There are many examples of developmental plasticity and adaptability in the literature. In this Section we shall describe three examples, each illustrating a different aspect of the general phenomenon.

9.1.1 The development of bone

That the process of development is highly coordinated and integrated is obvious to anybody with even a passing acquaintance with biology. But what is not often realized is the extent to which function becomes a mechanical cause of form. Adaptability, rather than being merely a reaction to altered circumstances, is essential to normal development. In other words, throughout normal development an organism must continually adjust to the functional demands set by its interaction with the environment; it is the sum of such interactions, at many levels, that determines the final phenotype. The development of vertebrate bone is a clear example of this principle.

The bony elements of the skeleton fit together in a precise fashion and the joints and articulations are well adapted to the functions which they serve. This is so not only in the adult animal but also early in development when great changes in shape and size are taking place.

Even the pattern in which the bony material is laid down inside the bone, changes. D'Arcy Thompson relates how the engineer, Professor Culmann of Zurich, happened in the year 1866 to come upon his biologist colleague Hermann Meyer when the latter was in the dissecting room. On seeing the arrangement of the lattice of bone in a section through the head of a femur exclaimed, 'That is my crane!' There seems indeed to be a resemblance between the lattice pattern in the head of a femur (Figure 30a) and a diagram of the lines of stress in the new and powerful crane that Culmann was involved in designing (Figure 30b). This strongly suggests that the structure of bone is finely attuned to the mechanical function it serves.

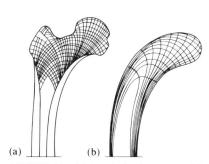

(a)　　　(b)

FIGURE 30 (a) Head of a femur, and (b) head of a crane, showing similar lines of tension and compression.

9.1.2 Metamorphosis in insects

During development, most insects pass through a succession of larval forms which precede the adult or imago stage. Each larval stage is referred to as an instar and is more or less distinct morphologically.

Metamorphosis in insects is subject to influences from a variety of environmental and social factors. Under certain conditions, paedomorphosis can occur: either progenesis in which reproductive maturity appears precociously, or neoteny in which somatic development is retarded so that the adult retains certain larval features. Wigglesworth (1954) suggested that temperature is a controlling factor in the metamorphosis of the large blood-sucking bug, *Rhodnius*. In a series of laboratory experiments, he obtained the result shown in Figure 31.

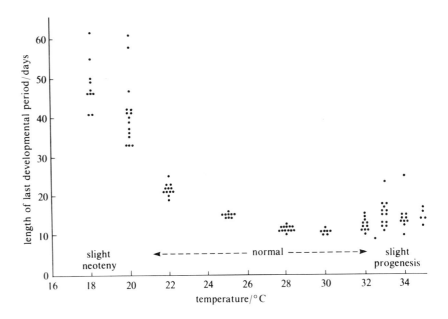

FIGURE 31 Effect of temperature on moulting of fourth stage larvae of *Rhodnius*. Moulting is delayed at low temperatures, with resulting neoteny. High temperatures speed development and produce progenesis.

ITQ 16   Which of the following statements are correct interpretations of the data in Figure 31?

(a) The rate of development decreases in general with increase in temperature.

(b) The rate of development increases in general with increase in temperature.

(c) There is a retardation of somatic development relative to reproductive maturity at low temperatures and an acceleration of reproductive maturity at high temperatures.

(d) There is an acceleration of reproductive maturity relative to somatic development at low temperatures and a retardation of somatic development at high temperatures.

It is of interest to note that development is 'normal' within a certain range of temperature (between 21.2 and 32 °C).

ITQ 17   What does this tell us about the stability of the process of development?

To explain the data, Wigglesworth suggested that the activity of juvenile hormone (which maintains the juvenile state) might be enhanced at low temperatures and that this would have the effect of slowing down insect development. To put it in more general terms, he suggested that an environmental factor (temperature) could influence development (via the chemical and physiological effects of a hormone). Some indirect evidence gives substance to this suggestion. In aphids, where there is a polymorphism of winged and wingless forms, it has been shown that juvenile hormone suppresses the growth of wings and that, further, winglessness is a juvenile condition. High levels of juvenile hormone also seem to be responsible for the permanently juvenile state of some salamanders (amphibians belonging to the order Urodela, see Unit 6 and the television programme 'Time for change').

9.1.3 Leaf development in *Ranunculus*

The development of many plants involves more or less abrupt changes in morphology. For example in a subgenus of *Ranunculus*, there are three groups of species of water crowfoots which are distinguishable by their leaf shapes. A typical member of the first group is the amphibious species *R. peltatus*, which inhabits small ponds in which there is considerable variation in water level. In this species, the first leaves are submerged and finely dissected, whereas later leaves float at the air–water interface and are flat and entire (Figure 32a). Experiments have shown that the change from dissected to entire leaves is also dependent on the length of the light–dark cycle to which the plant is subject. In other words, the environmental conditions act as a switch or trigger for some internally regulated alternative pathways of development that lead to widely different morphological results (varying leaf-shapes).

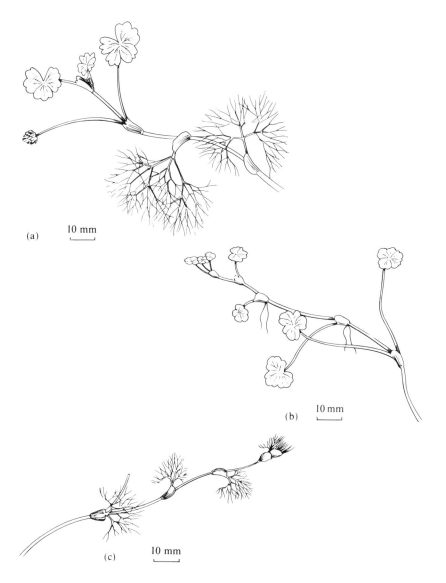

(a) 10 mm

(b) 10 mm

(c) 10 mm

FIGURE 32 Three species of water crowfoot that inhabit different environments. (a) *Ranunculs peltatus*. (b) *R. omiophyllus*. (c) *R. trichophyllus*.

A second group represented by *R. omiophyllus* (round-leaved water crowfoot) inhabits wet terrestrial habitats and shallow waters, and produces only entire leaves (Figure 32b); no manipulation of the environment will make these plants produce dissected leaves. Conversely, the third group, represented by *R. trichophyllus* (short-leaved water crowfoot) grows permanently submerged in deeper waters and produces dissected leaves only (Figure 32c).

In this example, phenotypic plasticity is exhibited in one group of species inhabiting a varying environment, while there is a lack of plasticity in related groups of species in more constant environments. This suggests the following evolutionary sequence:

1   Alternative phenotypes are generated initially as a developmental response to differing environmental conditions.

2　These alternative phenotypes become genetically fixed in certain species (see Sections 9.2 and 9.3).

In fact, the phylogeny worked out for *Ranunculus* based on anatomical, ecological and cytological considerations is compatible with this sequence: the groups typified by *R. omiophyllus* and *R. trichophyllus* are thought to have evolved more recently than the *R. peltatus* group.

9.2　The Baldwin–Morgan effect

A new idea, now referred to as organic selection, occurred more or less simultaneously to the Victorian biologist/philosophers J. M. Baldwin and L. Morgan. The gist of it was summarized by Baldwin in 1896. It involves a sequence of events which lead to evolutionary change:

1　Organisms developing in a new environment undergo adaptive modifications which enable them to survive in that environment.

2　Genetic mutations arise which give modification in the same direction. These are referred to as 'coincident variations'.

coincident variations

3　Natural selection acts to preserve the coincident variations, and evolution will proceed in the direction originally marked out by adaptive modification.

□ In what way does this sequence of events differ from the usual account of adaptation by natural selection given by neo-Darwinism?

■ It differs in the emphasis it places on the developmental adaptations of individual organisms. It is the specific and perhaps adaptive response of organisms to new environmental stimuli which *directs* evolutionary change.

In other words, developmental change is no longer regarded as something separate from the environment and arising from random gene mutations alone.

This idea allows for the fact that interactions between organisms and environment can sometimes produce adaptive modifications which are then passed on through the generations. This could be taken to imply that characters acquired by parents are inherited by offspring, a theory first proposed by Jean Baptiste de Lamarck in the early nineteenth century and long out of favour. To avoid Lamarckian associations, Baldwin and Morgan postulated the existence of 'coincident variations'.

inheritance of acquired characters

□ What makes the natural selection of coincident variants difficult to accept?

■ If we remember that selection must act via the phenotype, then it cannot effectively distinguish individuals with adaptive modifications alone from others carrying coincident variations; by definition, the two phenotypes would be exactly the same.

This raises once again the whole issue of how adaptations actually arise in evolution, and what precisely does the inheritance of acquired characters entail. We shall now turn to a discussion of these problems.

9.3　The epigenetic landscape

In a book that is now a classic, Waddington (1957) addressed himself to the problem of adaptation in the context of modern genetics. One of the examples he chose by way of illustration was the formation of calluses on the human feet. If genes are responsible for calluses, how is it, he asked, that they are formed so specifically on the parts of the body that require them for protection, and not anywhere else? To explain this, Waddington said, we must suppose that any part of our skin is potentially capable of forming calluses and that those on the soles of our feet are formed by increased pressure during development. Yet the fact that these calluses are there in the embryo before the feet have ever touched the ground, indicates that they must be produced by hereditary mechanisms independently of the environmental stimulus to which they are an adaptation. How did this come about?

Suppose that in an ancestral population of hominids or prehominids, the soles of the feet thickened in response to increased pressure when the bipedal gait was being evolved. As the response was adaptive, Waddington argued that it would become more and more pronounced, and would appear earlier and earlier in development; at the same time it would become regulated, so that essentially the same response would result from a range of intensities of the environmental stimulus. This he supposed to be accomplished by the selection of modifier genes, or genes which exert small modifying influences on the main response. The response would be *canalized*, that is, buffered against environmental and genetic disturbances. Meanwhile, certain gene mutations occurring at random could 'fix' this response in the developmental process, so that it happens in the *absence* of the environmental stimulus. The latter process is called *genetic assimilation*. Thus, the combination of developmental adaptation, canalization and genetic assimilation are said to 'mimic' Lamarckian inheritance.

modifier genes
canalization

genetic assimilation

There is perhaps no formal difference between Waddington's idea and that of Baldwin and Morgan, except that Waddington called attention to certain properties of the developing system which he summarized as follows:

1 Developmental processes are in general buffered against both environmental and genetic disturbances; in other words, such processes are canalized to produce essentially the same phenotype. This is developmental homeostasis (see Section 9).

2 Development may proceed along alternative pathways to give different end results. This is developmental plasticity or adaptability (see Sections 9, 9.1 and 9.3).

3 Complex relationships between genotypes and phenotypes are the rule rather than the exception (see Section 9.1).

Waddington pictured the developing system as a landscape with hills and valleys, the geographic features of which are determined by 'forces' exerted by various gene products which interact with each other; he referred to it as an epigenetic landscape. The developmental paths of tissues and cells are thus constrained to 'flow' along certain valleys and not others, and various paths will be marked out and separated from potential branch points by suitable hills which confer on the process its homeostasis. Because the landscape is a dynamic structure capable of being remoulded by both genetic and environmental influences, other paths are potentially accessible to give different end results. This accounts for developmental plasticity. And because the system reacts as a whole to disturbances, it is not surprising that genetic mutations at more than one locus may often give the same results (see Figure 33).

epigenetic landscape

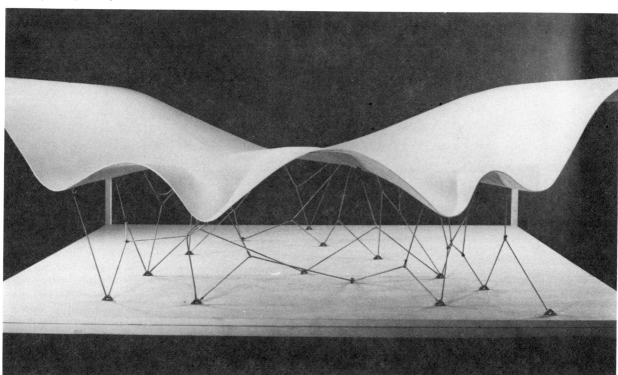

FIGURE 33 Waddington's epigenetic landscape—the complex system of interactions underlying the development of organisms. The pegs in the ground represent genes; the strings leading from them, the chemical tendencies which the genes produce. The moulding of the epigenetic landscape is controlled by the pull of these numerous guyropes which are anchored to the genes.

The epigenetic landscape puts much emphasis on the whole system of interacting genes and gene products which becomes the unit of natural selection. If this is indeed what happens, a primary object of evolutionary studies must be those principles in internal organization of the epigenetic system that permit survival and are responsible for changes in development.

A number of experiments have been designed to test the ideas of canalization and genetic assimilation. In one of these, Waddington demonstrated the genetic assimilation in *Drosophila* of the *bithorax* phenotype (see Figure 34). The appendages called balancers or halteres become transformed, partially or wholly, into wings.

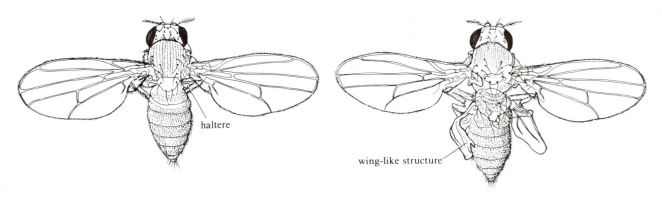

FIGURE 34 *Drosophila*. (a) A normal fly; (b) a bithorax phenocopy.

There is a genetic mutation which gives rise to this phenotype, but mimics of the mutant phenotype, called *phenocopies*, can be induced in normal strains of flies by treating the eggs with ether fumes. Waddington exposed a large number of eggs to ether and examined the emerging adults for bithorax phenocopies. In each of two experiments, two lines were started using the offspring of these flies, one in which only bithorax flies were mated together (upward selection) and the other in which only flies with normal phenotypes were mated together (downward selection). Ether treatment of the eggs was done in every generation in all lines, and the process of selection repeated. The results are shown in Figure 35. In addition, some eggs in each generation were allowed to hatch without treatment.

phenocopies

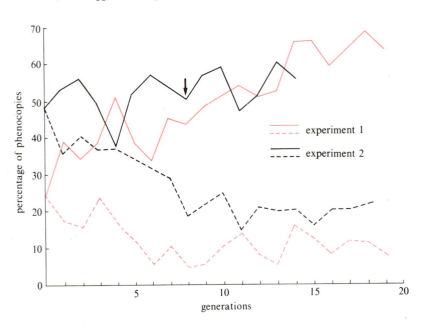

FIGURE 35 Percentage of bithorax phenocopies in successive generations of upward (plain lines) and downward (dashed lines) selection following ether treatment of the eggs. The arrow at generation 8 in experiment 2 indicates the point at which the first assimilated stocks appeared.

ITQ 18 There is a major flaw in this experiment; can you work out what it is?

Upward selection appears to have been quite effective in increasing the frequency of the bithorax phenocopy, while downward selection reduced but did not eliminate the phenocopy, even after 19 generations.

After a number of generations of upward selection, it was observed that some bithorax adults appeared even without ether treatment. By breeding these adults together and thus selecting for spontaneous appearance of bithorax, true breeding bithorax lines were obtained which consistently gave 70 to 80 per cent of bithorax offspring. That is, the phenotype had become genetically assimilated. In two of these lines a single dominant mutation seemed to be responsible, whereas in the third a polygenic system was involved.

Do these results lend support to the idea of the epigenetic landscape? The answer is yes. First, the existence of phenocopies is in itself a very important phenomenon. Bithorax is only one of a very large number of phenocopies induced by environmental stimuli, each of which mimics the action of a different mutant gene. This means that it is the internal organization of the epigenetic system, and not just individual genes, which determines the types of change that can occur. Second, the fact that more than one kind of genetic mechanism can give rise to the same phenotype is again indicative of a complex network of relationships between genes, rather like the intuitive picture suggested by the idea of the epigenetic landscape.

One major criticism of Waddington's experiment is that the bithorax phenotype does not appear to be an adaptation to ether treatment, and therefore the experiment tells us very little about how real adaptations arise.

In another experiment, three strains of *Drosophila melanogaster* were grown on medium to which sodium chloride had been added in sufficient quantities to cause the death of over 60 per cent of the larvae. By increasing the salt concentration in the medium for later generations, the intensity of selection was maintained at a roughly constant level for 21 generations. At that stage, known numbers of eggs were set up in media with various salt concentrations in order to test for larval resistance to salt. Unselected stocks like those used at the start of the experiment were tested simultaneously, as controls. The survival rate on each medium was ascertained for each group, and measurements were made of the area of the anal papillae—organs thought to be involved in osmotic regulation (i.e. control of solute concentrations in body fluids). The results are shown in Figure 36.

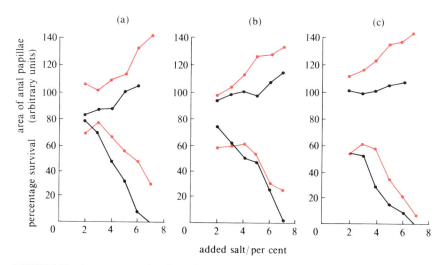

FIGURE 36   Percentage survival (below) and size of anal papillae (above) in three different selected (red) and unselected (black) strains in various concentrations of salt. The selected strains had been subjected to the indicated concentrations of salt for 21 generations.

The results are clear cut: the strains subjected to sodium chloride for 21 generations are better adapted to sodium chloride than the control strains subjected to it for the first time, and have larger anal papillae. The relevance of these results to the model of developmental adaptation, canalization and genetic assimilation presented earlier is unclear, however, because:

1   The results do not distinguish between the effects of environmental selection and the effects of developmental response to environmental stress. In particular, the increase in size of the anal papillae in high concentrations of sodium chloride is a characteristic of the unselected strains in Figures 36(a) and (b) and to a minor extent in Figure 36(c); therefore the selection of alleles giving larger papillae may have occurred in the experiment without there having been any change in the developmental response as such.

2   The results do not give any clear indication that canalization has occurred. As can be seen in Figure 36, the size of the anal papillae increases fairly steadily with increase in sodium chloride concentration for the selected strains. If canalization had occurred, then approximately the same size of anal papillae would have been found over a certain *range* of sodium chloride concentrations.

3   Although one could claim some evidence of genetic assimilation, in the sense that the size of the anal papillae are enlarged in the selected strains even at the lowest concentration of sodium chloride, we cannot legitimately say that genetic assimilation has taken place because we cannot distinguish the effects of selection from the effects of canalization of a novel response (see 1 above).

What has been demonstrated clearly by experiments such as these is that genes and gene-combinations giving the required morphological adaptations are far too common to be accounted for by specific alleles at only a few loci. Genetic analyses indicate that assimilated characters are usually controlled by many genes which are present in *all* chromosomes. This speaks strongly in favour of the need to consider systems of genes or the development of whole organisms in the explanation of evolution. The so-called epigenetic approach is increasingly being adopted by those evolutionists who believe that the key to understanding evolution lies in the mechanisms of development and in how these mechanisms interact with the environment to produce evolutionary change.

9.4   The inheritance of acquired characters

Finally, it remains to say a few words on the subject of the inheritance of acquired characters. Lamarck's original formulation was simply a statement about a phenomenon and not a theory about the mechanism of heredity. Moreover, contrary to many accounts in textbooks and the scientific literature, Lamarck did *not* assert that the adaptations or modifications acquired are passed on to the next generations through the genes or the DNA. Like Darwin, he knew nothing about the mechanisms of heredity that we are familiar with today. Instead, Lamarck (1809) saw the phenomenon occurring as follows:

> ... when individuals of any species change their situation, climate, mode of existence or habits, their structure, form organisation, and in fact their whole being becomes little by little modified till in the course of time it responds to the change experienced by the creature.

> In the same climate difference in situation, and a greater or less degree of exposure, affect simply, in the first instance, the individuals exposed to them, but in the course of time, these repeated differences of surroundings in individuals which reproduce themselves continually under similar circumstances induce differences which become part of their very nature; so that after many successive generations, these individuals, which were originally, we will say, of any given species, become transformed into another one.

Thus, Lamarck clearly saw the permanent alterations required for transformation of species as taking place after many successive generations, during which the change in environment persists. The inheritance of acquired characters is quite irrelevant to his scheme. Because of the persistence of the change in environment, the same modification would arise in development without the need to postulate any genetic modification initially. The whole phenomenon bears a remarkable resemblance to the processes of developmental modification, canalization and genetic assimilation described in the previous Section. The only difference is that, a century and a half later, we are able to see a means whereby the modification could become genetically 'fixed', possibly through random mutations. It is time we stopped discrediting Lamarck unjustly and recognized that his whole emphasis on organism–environmental interactions in the origin of adaptive variations is a very fertile framework for further evolutionary research.

9.5   Objectives and SAQs for Section 9

Now that you have completed this Section, you should be able to:

(a)   Define and recognize definitions of the terms developmental homeostasis, developmental and phenotypic plasticity, coincident variation, inheritance of acquired characters, canalization and genetic assimilation.

(b)   State in general terms the relationship between genes and phenotype.

(c) Evaluate, criticize and suggest experiments on the mechanisms of canalization and genetic assimilation of novel phenotypes.

To test your understanding of Section 9, try the following SAQs.

SAQ 10 (*Objectives a and c*) In the experiment to demonstrate the genetic assimilation of the bithorax phenotype (see Section 9.3), which of the following results, singly or in combination, are of primary importance in the demonstration of (a) canalization, and (b) genetic assimilation?

(i) The frequency of bithorax in the starting stock of flies is zero per cent.

(ii) The frequency of bithorax in each generation is higher than that in the previous generation.

(iii) Bithorax phenotypes appear in emerging adults without ether treatment of the eggs in later generations.

(iv) The bithorax phenotypes in later generations are very uniform in appearance.

(v) The bithorax phenotypes in early generations are variable in appearance.

SAQ 11 (*Objectives b and c*) In the bithorax experiment, two competing hypotheses were advanced to explain the phenomenon of genetic assimilation:

(i) that a chance mutation in a single gene is involved which substitutes for the role of the environmental stimulus;

(ii) that selection of a number of modifier genes is involved, which makes the response to the environmental stimulus more and more pronounced so that the response eventually comes about without the environmental stimulus.

Which of the hypotheses is supported by the evidence described in Section 9.3?

SAQ 12 (*Objective c*) An evolutionist claims that the effect of ether treatment could be transmitted via the egg cytoplasm from one generation to the next. This causes the response to deepen in successive generations of ether treatment. He believes that such a mechanism, rather than the selection of modifier genes, is responsible for the canalization and genetic assimilation of the bithorax phenotype. How would you modify the experimental design in order to test this hypothesis? Describe a breeding test you would carry out to ascertain whether some form of heredity resides in the egg cytoplasm.

10 Conclusion

This Unit has been about macroevolution: large or discontinuous morphological changes in evolution. We began by reviewing the various genetic mechanisms that have been postulated to give rise to abrupt changes in evolution. Without an understanding of how genes determine morphology, however, any theory of evolution must remain incomplete.

Next we dealt with the relationship between macroevolutionary change and development—the process by which form is generated. We argued that ontogenetic modifications are the basis of phylogenetic change. If genes affect the rates of different reactions in development, for example, then a mutation causing a change in the velocity of one reaction would alter its rate relative to other reactions, resulting in large physiological and phenotypic effects. This alteration in the time-coupling between various component processes in development is referred to as heterochrony. Heterochrony can lead to large changes in adult morphology. As the fossil record typically consists of a sequence of adults, such changes would appear in an exaggerated form, giving the impression of big gaps in the fossil record.

However, heterochrony not only results from gene mutations, but can also be induced by environmental factors. This brought us to an examination of the possible role of the environment in the initiation of evolutionary change. The environment acts not merely as the agent of natural selection, but also as the medium with which the developing organism interacts in generating novel morphological responses. The mechanisms whereby such responses may become fixed in the heredity of the organisms were then discussed.

At the time of writing of this Unit (1981) there is a growing interest in many possible mechanisms of evolution in addition to natural selection. We have touched upon some of these in the Unit. For example, the nature of developmental laws and constraints may limit the number of *possible* phenotypes before natural selection could be said to act. Organism–environmental interactions may be responsible for initiating at least some morphological and physiological changes in evolution. The nature of these interactions and, subsequently, the mechanisms whereby phenotypic changes could become genetically assimilated are being investigated. In particular, there appears to be the possibility that instruction could pass directly from the environment to the heredity apparatus, at least under certain circumstances.

Rapid advances in molecular biology increasingly draw attention to the enormous complexity of genetic organization. The simplistic concept of 'gene' in neo-Darwinian theory corresponds to nothing in molecular terms. There is little doubt, in the minds of many population geneticists, that the genetic theory of evolution itself will be subject to drastic reformulation to take into account all the new information in molecular genetics that has already accumulated.

One thing is clear: evolution is a very complex phenomenon and demands a much broader explanation than that offered by the natural selection of random mutations.

General references

BALDWIN, J.M. (1896) A new factor in evolution, *American Naturalist*, **30**, 441–51.

BRITTEN, R.J. and DAVIDSON, E.H. (1969) Gene regulation for higher cells: a theory, *Science*, **165**, 349–357.

BRITTEN, R.J. and KOHNE, D.E. (1968) Repeated sequences in DNA, *Science*, **161**, 529–540.

DAVIDSON, E.H. and BRITTEN, R.J. (1979) Regulation of gene expression: possible role of repetitive sequences, *Science*, **204**, 1052–9.

GOULD, S.J. (1977) *Ontogeny and phylogeny*, Belknap Press, Harvard University.

HUXLEY, J. (1932) *Problems of Relative Growth*, Methuen.

LAMARCK, J.B. (1809), *Zoological Philosophy*, cited in BUTLER, S. (1911) *Evolution, Old and New*, Fifield.

MAYNARD SMITH, J. (1969) *Towards a Theoretical Biology 2: Sketches*, ed. C.H. Waddington, Edinburgh University Press.

SCHINDEWOLF, O.H. (1936) *Paleontologie, Entwicklungs: Lehre und Genetik*, Bornträger, Berlin.

THOMPSON, D'ARCY, W. (1917) *On Growth and Form*, Cambridge University Press.

WADDINGTON, C.H. (1957) *The Strategy of the Genes*, Allen and Unwin.

WIGGLESWORTH, V. (1954) *The Physiology of Insect Metamorphosis*, Cambridge University Press.

Further reading

General

BUTLER, S. (1911) *Evolution, Old and New*, Fifield.

FRAZZETTA, T.H. (1975) *Complex Adaptations in Evolving Populations*, Sinauer Associates Inc.

GOULD, S.J. (1977) *Ontogeny and Phylogeny*, Belknap Press, Harvard University.

THOMPSON, D'ARCY, W. (1961) *On Growth and Form*, ed. J.T. Bonner, Cambridge University Press.

WADDINGTON, C.H. (1957) *The Strategy of the Genes*, Allen and Unwin.

Alternatives to neo-Darwinian theory

(a) Critiques of neo-Darwinism

CANNON, H.G. (1959) *Lamarck and Modern Genetics*, Manchester University Press.

LØVTRUP, S. (1976) On the falsificability of neo-Darwinism, *Evolutionary Theory*, **1**, 267–283.

MACBETH, N. (1974) *Darwin Retired*, Garnstone Press.

MOORHEAD, P.S. and KAPLAN, M.M. (1976) *Mathematical Challenge to the neo-Darwinian Interpretation of Evolution*, Wistar Inst. Press.

WHYTE, L.L. (1965) *Internal Factors in Evolution*, George Braziller.

(b) Epigenetic or developmental approaches to evolution

ALBERCH, P. (1980) Ontogenesis and morphological diversification, *American Zoologist*, **20**, 653–667.

HO, M.W. and SAUNDERS, P.T. (1979) Beyond neo-Darwinism—an epigenetic approach to evolution, *Journal of Theoretical Biology*, **78**, 573–591.

SAUNDERS, P.T. and HO, M.W. (1981) On the increase in complexity in evolution II: the relativity of complexity and the principle of minimum increase, *Journal of Theoretical Biology*, **90**, 515–530.

WEBSTER, G. and GOODWIN, B.C. (1981) Rethinking the origin of species by natural selection, in *Against Biological Determination: Towards a Liberatory Biology.* ed. M. Barker, *et al.*, Allison and Busby.

(c) Neo-Lamarckian theories of inheritance and evolution

COOK, N.D. (1977) The case for reverse translation, *Journal of Theretical Biology*, **64**, 113–135.

HO, M.W. and SAUNDERS, P.T. (1981) Adaptation and natural selection: mechanism and teleology, in *Against Biological Determination: Towards a Liberatory Biology*, ed. M. Barker *et al.*, Allison and Busby.

KOESTLER, A. (1971) *The Case of the Mid-wife Toad*, The Anchor Press.

ROSEN, D.E. and BUTH, D.G. (1980) Empirical evolutionary research versus neo-Darwinian speculation, *Systematic Zoology*, **29**, 300–308.

STEELE, E.J. (1979) *Somatic Selection and Adaptive Evolution: on the inheritance of acquired characteristics*, Williams and Wallace International Inc.

ITQ answers and comments

ITQ 1 (i) and (iv) are microevolutionary, and (ii), (iii) and (v) are macroevolutionary phenomena.

ITQ 2 The rate at which chronospecies arise in a lineage is many times slower than that at which new species are formed by the splitting of lineages. This suggests that different processes may be involved in the gradual transformation of a single lineage (similar to microevolution) and in the explosive adaptive radiation by branching or punctuated speciation (macroevolution).

ITQ 3 The genetic processes postulated to give large changes are:

(i) founder effects in small populations;

(ii) inbreeding in small populations, leading to increased homozygosity;

(iii) genetic drift leading to fixation of alleles;

(iv) organismic selection for the alleles that function best in homozygous combinations.

The result of these processes is a drastic reorganization of the genome, or genetic revolution (see Unit 12, Section 5.1.2).

ITQ 4 (a) (i) In Figure 4(a) the repeated elements are the receptor genes, R_1, R_2 and R_3; (ii) in Figure 4(b) they are the integrator genes, I_A, I_B, I_C and I_D.

(b) The activated producer genes would be (i) P_A and P_C in Figure 4(a), and (ii) P_B, P_C and P_D in Figure 4(b).

ITQ 5 On anatomical and fossil evidence, the mammals—a class which includes the multituberculates, the insectivorous mammals, the true mammals (marsupials and placentals) and the monotremes—are all believed to be derived independently from different groups of therapsid reptiles (see Unit 6).

(ii) The holostean fishes (see Unit 5).

ITQ 6 (i) The placental and marsupial radiations, which occurred independently—the latter in Australia, the former in other continents (see Unit 8, and the television programme 'Continental arks').

(ii) The iterative evolution of the planktonic foraminifera. Parallel radiation was repeated in different epochs, with a mass extinction of all but one form at the end of the Eocene (see Unit 8, Section 2.3).

ITQ 7 If the length of the elephant is fifty times that of the mouse, its weight is 50^3 or 125 000 times that of the mouse. In order to support its huge weight, the cross-sectional area of the elephant's leg will have to be 125 000 times that of the mouse. Had the shape remained the same, the 'elephant-sized mouse' would have had legs only $50^2 = 2 500$ times the cross-sectional area of the normal-sized mouse leg.

ITQ 8 Multicellular organisms all start life as a single cell; therefore differential growth is necessary for them to attain adult shape and size.

ITQ 9 The growth of antlers in deer, culminating in the enormous antlers of the Irish elk (Unit 8, Section 2.4).

ITQ 10 (a) $$s = \left(\frac{b_d}{b_a}\right)^{1/1-k} = \left(\frac{0.210}{0.237}\right)^{-1/0.766} = 1.17$$

(b) It means that the descendant has the same shape as the ancestor when it is 1.17 times as large, that is, shape is slightly retarded with respect to body size.

ITQ 11 (a) $\log p_a = \log 0.237 + 1.766 \log r_a$; $\log p_d = \log 0.210 + 1.766 \log r_d$. (b) This means that the slopes (k) of the two curves are equal.

ITQ 12 There are two kinds of paedomorphosis: progenesis (the acceleration of maturation) and neoteny (the retardation of somatic development).

ITQ 13 The term r-selection refers to the selective effects of density-independent factors in conditions where the populations increase continuously without being held back by the limited resources available in the environment. Under such conditions, there is selection for increased population growth rate r. The term K-selection refers to the effects of density-dependent factors in conditions where the population remains at or near its saturation density, seldom undergoing any prolonged periods of increase. Selection here is thought to result in an increase in the carrying capacity K of the environment for the particular organism, accompanied by an increased ability of individuals to compete with other organisms and to utilize the limited resources.

ITQ 14 Progenesis might be favoured by unstable environments or environments subject to periodic catastrophes that tend to keep the population down. Neoteny might be associated with stable environments where population density is at or near saturation.

ITQ 15 Limpets inhabiting brackish water have bumpy shells, whereas limpets inhabiting fresh water have smooth shells but are otherwise identical to those found in brackish water. When a limpet is transplanted from fresh to brackish water, the new growth at the margin of the shell is bumpy. This suggests that it is the same species that inhabits both brackish and fresh water, showing different morphologies in each environment.

ITQ 16 (a) Incorrect. (b) Correct. This explains the appearance of neoteny at low temperatures and progenesis at higher temperatures. (c) Correct. (d) Incorrect.

ITQ 17 There is developmental homeostasis—a buffering against changes within a certain range of environmental temperatures.

ITQ 18 There is no control in this experiment. In other words, a third line which involved *no selection* should have been started at the same time, so that changes independent of selection could have been distinguished.

SAQ answers and comments

SAQ 1 The major difficulties are as follows:

(i) Macroevolution involves large morphological changes of a kind never observed in microevolutionary phenomena.

(ii) The postulated intermediates of macroevolution are seldom found among either extant or extinct groups of organisms.

(iii) The fossil record supports a general picture of 'punctuated equilibria' which is inconsistent with *continuous* small changes.

(iv) The lack of an understanding of the relationship between genes and morphology means that the current theory, based solely on the changes in allele frequencies in populations, does not address the question of macroevolution at all (see Section 2).

SAQ 2 (a) S_1 or S_2, i.e. the genes linked to integrator genes I_1 and I_2 which, in turn, induce receptor genes R_1 and R_2.

(b) P_B and P_D, i.e. those genes activated via R_B/I_B and R_D/I_D. (See Section 4.2.)

SAQ 3 The Davidson–Britten models tell us nothing about the mechanisms responsible for the differential activation of sensor genes. Moreover, they have nothing to say about the spatial patterns of tissues and organs, or the temporal order in which different genes are turned on in different parts of an organism.

SAQ 4 The observed parallels between the developmental stages of an individual and the phylogenetic sequence of its adult ancestors gave rise to Haeckel's law of recapitulation. Von Baer, however, observed from embryological studies that the *young stages* of different mammals often resembled one another more than the adult stages. This led him to his laws of development.

SAQ 5 Haeckel's law is explicitly evolutionary, because it explains the resemblances between embryos and their adult ancestors as due simply to descent with modification: in other words, evolution. Von Baer's laws, on the other hand, explain the resemblances between different embryos on the basis of universal laws of development. There is no need for heredity or descent, and hence no need for an evolutionary explanation.

SAQ 6 (a) False. The solid line represents an interspecific allometry.

(b) False. Mammals in general have smaller brain weights *in proportion to their body weights* than primates.

(c) True.

(d) False.

(e) True. This can be deduced by comparing the positions of *Australopithecus* (A), *Homo erectus* (E) and *Homo sapiens* (S).

SAQ 7 (a)

For species A: $\log b = 0.1, b = 1.26, k = 1.2, \log c_A = 0.1 + 1.2 \log l_A$
For species B: $\log b = 0.1, k = 1.4, \log c_B = 0.1 + 1.4 \log l_B$

The constant 0.1 is the intercept on the log c axis, and the slopes 1.2 and 1.4 are determined visually from the graphs.

(b) The single parameter that has changed is the allometric coefficient k (from 1.2 in A to 1.4 in B).

(c) The values listed in Table 5 can be determined directly from the graphs using antilog tables or a suitable pocket calculator.

TABLE 5

	size, l/mm	shape, c/l
species A	31.62	2.51
species B	17.78	3.98

(d) To determine whether the shape of B is retarded or accelerated with respect to A, we really want to know if the shape A has at a particular stage of development is attained by B later (at larger size) or earlier (at smaller size). For the comparison we choose a standard age, that of sexual maturity. Through the point indicated by the arrow on the graph of A (the point where sexual maturity is reached) draw a line of slope 1 (the broken line through the point P in Figure 37). This intersects the graph of B at a point Q, which corresponds to a considerably smaller size than point P. Hence, shape is accelerated in species B compared with species A. (Re-read Section 8 if you had difficulties with this question.)

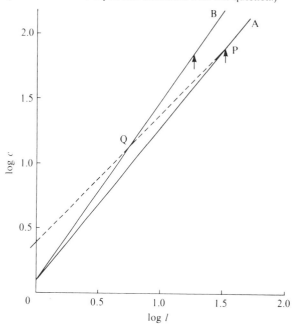

FIGURE 37 Method for determining whether shape is retarded or accelerated in the descendent species.

(e) The clock diagram would look like Figure 38. The net result is recapitulation. As we have seen in (d), the shape in species B exceeds that in species A. The rate of sexual development remains the same, since they both take 200 days to reach maturity. Somatic development in size is retarded, however, since after the 200 days (at the points marked with arrows) l is smaller for B than for A. This is clearly not one of the 'pure' forms of heterochrony shown in Figure 26.

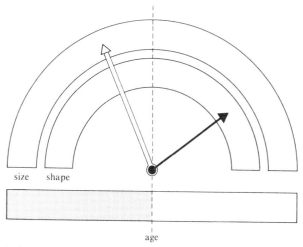

FIGURE 38 Clock model for heterochrony in the evolution of species B from species A.

SAQ 8 The genes involved would be those controlling the rate of growth of body circumference. The change in species B is an *increase* in that rate of growth.

SAQ 9 There are two similar ontogentic phases of brain growth in all three primate species: a rapid early phase, followed by a slower phase. Comparing human ontogeny with that of the other species, the main difference is a prolongation of the early phase of brain growth, and this is the most probable explanation for the larger brain in humans.

SAQ 10 (a) (i), (iv) and (v); (b) (i) and (iii). Result (i) is crucial in both cases, in order to exclude the presence of pre-existing mutant genes giving the bithorax phenotype. That is, the phenotype must result from the environmental stimulus and not have been present already. Results (iv) and (v) are necessary in (a) because canalization involves just that: a uniform response arising in a range of intensities of the stimulus.

SAQ 11 The results show that *both* hypotheses could be correct. In two of the lines a single gene mutation has been identified, whereas in the third line many genes, possibly modifiers, are involved.

SAQ 12 The experiments should include a control line which is ether treated in every generation, as in the upward- and downward-selection lines, but *without* selection. An increase in the frequency of bithorax phenotypes in the control lines would support the hypothesis. One breeding test which could be done is to bring about matings between the treated lines and an untreated stock to see if the progeny resulting from the cross (treated female × untreated male) gives higher frequencies of bithorax than that resulting from the reciprocal cross (untreated female × treated male).

S364 Evolution

Acknowledgements

Grateful acknowledgement is made to the following for permission to reproduce Figures in this Unit:

Figure 1 from Britten, R. J. and Davidson, E. H. (1969) 'Gene regulation for higher cells: a theory' in *Science*, vol. 165, pp. 349–357. Copyright © 1969 by the American Association for the Advancement of Science; *Figure 2* from Britten, R. J. and Kohne, D. E. (1968) 'Repeated sequences in DNA' in *Science* vol. 161, pp. 529–540. Copyright © 1968 by the American Association for the Advancement of Science; *Figures 10, 12, 14* from Huxley, J. (1932) *Problems of Relative Growth*, Methuen and Co. Reprinted by permission of Anthony Huxley; *Figure 17* from Szalay, S. (1975) 'Approaches to primate paleobiology' *Contrib. Primat.* vol. 5, S. Karger AG, Basel; *Figures 18 and 19* from Goldschmidt, R. (1934) in *Bibliograph Genetica*, vol. 11; *Figures 29, 25, 26, and 27* from Gould, S. J. (1977) *Ontogeny and Phylogeny*, Belknap Press, a division of Harvard University Press; *Figure 31* from Wigglesworth, V. (1954) *The Physiology of Insect Metamorphosis*, Cambridge University Press; *Figure 35* from Wassington, C. H. (1956) in *Evolution*, vol. 10, Society for the Study of Evolution, University of Georgia; *Figure 36* from Waddington, C. H. (1959) 'Canalisation of development and genetic assimilation of acquired characters' in *Nature*, vol. 183, no. 1654. Copyright © 1959 Macmillan Journals Ltd.